AF686677

The Mathematics of Food

Dr. Edwin Masimba Moyo

ISBN: Number
Paperback: 978-1-971141-73-2
Hardcover: 978-1-971141-74-9

Published by:

Ultra Publishers

www.ultrapublishers.com

Printed in the United States of America

Dedication

This book is dedicated to all who strive to leave the world better than they found it – champions of humanity, climate resilience, and environmental restoration. To those fighting for a future where food nourishes, ecosystems thrive, and our shared humanity flourishes.

Acknowledgment

I acknowledge the guidance and inspiration of God, through Jesus Christ and the Holy Spirit, in writing this book. As reminded in 1 Corinthians 6:19-20, our bodies are temples of the Holy Spirit – a truth that underscores the importance of nurturing and honoring them through mindful choices about food and health.

About the Author

Dr. Edwin Masimba Moyo is an author and entrepreneur with a track record of successful agricultural projects in Zimbabwe and the region. He's the founder of Kondozi, a farming enterprise that employed 5,000 people and exported $18M worth of produce to UK supermarkets like Tesco. Dr. Moyo pioneered the blueberry industry in Zimbabwe, opening up Chinese markets and empowering small-scale farmers.

He's the promoter of AgriConnect Africa, a digital platform connecting farmers to markets and financial services. Dr. Moyo's mission is to fight poverty, hunger, and disease, and promote sustainable eating habits. His vision is to leave a better world than he found it, believing Africa can be an economic powerhouse by supporting small-scale farmers and businesses, and fostering intra-African and international trade.

To every man there comes a time in his lifetime that special moment when he is figuratively tapped on the shoulder and offered a chance to do a very special thing, unique to him and fitted to his talents. What a tragedy if that moment finds him unprepared or unqualified for the work which would be his finest hour

- Winston Churchill

Table of Contents

Foreword

The Mathematics of Food explores the intricate relationships between mathematics, food systems, and sustainability. This book delves into the complex world of food economics, tracing the path from production to consumption, and highlighting the critical role mathematics plays in shaping our food future.

Through its pages, you'll discover how mathematical models optimize crop yields, streamline supply chains, and predict food demand. You'll see how data-driven approaches inform policy decisions, promote sustainable agriculture, and ensure food security.

Whether you're a mathematician, food enthusiast, or concerned citizen, this book offers a fascinating journey into the hidden patterns and structures governing our food systems. Join us as we unravel the threads of mathematics, food, and sustainability – and discover a more resilient food future.

Before *After*

Amanda Lamb
Food nutrition advocate

By Natasha S. Seth

When I first met Dr. Edwin Masimba Moyo, the conversation unfolded with an ease that comes from shared concern and shared curiosity. We talked about food waste, about how difficult it is for small and independent growers to navigate global supply chains, and about the steady drift of food cultures toward what is "fast," "cheap," and "easy," even as health outcomes worsen across the world. Dr. Moyo spoke about his work connecting growers across Africa to suppliers and grocery markets in Western economies—work rooted in the belief that food systems should elevate people, strengthen local economies, and reduce food insecurity rather than entrench it.

As the conversation progressed, we found common ground in a simple but powerful realisation. Beneath the visible challenges of distribution and access lies a deeper structure—one shaped by incentives, assumptions, and calculations that quietly govern how food moves, who benefits, and where costs are absorbed. We kept returning to the same insight: to change outcomes, we first need to understand the system that produces them. Food, we agreed, is best understood not as a collection of isolated issues, but as an interconnected ecosystem shaped by mathematics. That insight became the starting point for this book.

Food is rarely discussed as a system, even though it behaves like one. The choices that determine what we grow, how food is processed, how it moves through supply chains, and who ultimately has access are shaped by forces that operate largely out of sight. The Mathematics of Food: Numbers from Field to Plate brings those structures into view and uses them as a foundation for a more informed and constructive conversation about change.

The book examines the global food system, tracing the forces that shape outcomes at scale. It exposes the extent of loss created by inefficiency and waste, not only as a moral concern but as an economic one. It looks closely at the drivers behind food processing and additive use, and at how short-term commercial logic routinely shifts long-term costs onto public health systems. It also confronts the wider societal impact of unhealthy diets, showing how these outcomes accumulate across populations and generations.

Rather than treating these issues as isolated problems, The Mathematics of Food shows how they are deeply interconnected. Predictions of consumer behaviour and purchasing patterns—now increasingly shaped by AI-driven models—ripple through supply chains and surface later as environmental strain, rising healthcare costs, and food insecurity. The book uses models not to strip away complexity, but to make it legible, and to test alternative ways of organising food systems that prioritise health, resilience, and sustainability.

This book is written for anyone involved in shaping food systems, and for anyone living within them—not just specialists or economists. Its purpose is to replace slogans with clarity, and to move the conversation beyond good intentions toward informed choice. At a moment when pressure on food systems is intensifying from multiple directions, the question is no longer whether change is needed, but whether we are willing to examine the structures that produce current outcomes.

The Mathematics of Food argues that what we choose to measure and reward ultimately shapes what our food systems become. By making those choices visible across multiple, interlinked processes and decisions, the book invites readers to rethink how food is valued, governed, and sustained—and to engage more deliberately in shaping a food system fit for the future.

Preface

The Mathematics of Food: A Journey into the Numbers Behind What We Eat

Food is a fundamental part of human life, influencing our health, culture, and environment. Behind every meal, there's a complex web of numbers, patterns, and mathematical concepts that shape the way food is produced, processed, and consumed.

This book explores the fascinating intersection of mathematics and food systems, delving into topics like food economics, sustainability, and food security. We'll examine how mathematical models, data analysis, and computational tools help us understand and address pressing challenges in the food sector.

From optimizing crop yields to reducing food waste, mathematics plays a vital role in creating a more efficient, equitable, and sustainable food system. Through real-world examples, case studies, and accessible explanations, this book aims to illuminate the often-overlooked mathematical underpinnings of our food systems.

Whether you're a student, researcher, or simply curious about the mathematics behind food, this book invites you to explore the intriguing connections between numbers, food, and our global healthy food landscape.

– Dr. Edwin Masimba Moyo

Introduction: The Mathematics of Food

Food sits at the center of human existence. It sustains life, shapes societies, and determines how communities grow or fail. Mathematics gives us the tools to understand how food moves through the world, from the field to the plate. In this book, we explore how mathematical thinking quietly governs the way food is produced, processed, distributed, and consumed. At the core of this relationship lies one essential idea: **optimization**.

Optimization is the search for the best possible outcome within real limits. In food systems, those limits include land, water, labor, climate, cost, and time. Applying optimization allows us to make deliberate choices rather than reactive ones. Through techniques such as linear programming, statistical modeling, and game theory, mathematics helps improve crop yields, reduce waste, and strengthen supply chains. These tools are not abstract; they shape everyday decisions that affect food security, sustainability, and nutrition on a global scale.

Mathematics operates quietly behind modern food systems. It guides precision agriculture, informs planting schedules, and determines how resources are allocated across farms. It also powers the algorithms behind food delivery platforms and inventory systems that move products efficiently through cities. Understanding the mathematics of food allows individuals, producers, and policymakers to make informed decisions about production, distribution, and consumption. Over time, those decisions shape a food system that is more resilient, more equitable, and better prepared for future pressures.

This book examines the mathematical foundations of food systems across several key areas:

- Crop modeling and yield prediction
- Food processing and preservation
- Supply chain optimization
- Nutrition and meal planning
- Food waste reduction and management

Together, these topics reveal how numbers, models, and constraints influence what ends up on our tables. This is an invitation to look beyond intuition and tradition and to see food systems as dynamic structures that can be improved through careful analysis.

What Is the Mathematics of Food?

The mathematics of food is an interdisciplinary field that uses mathematical concepts and techniques to understand and analyze the systems involved in food production, processing, distribution, and consumption. These systems are complex, interconnected, and shaped by real-world constraints.

At its core, the mathematics of food focuses on using models and algorithms to examine how food systems function and how they can be improved. This includes questions of efficiency, sustainability, and equity, as well as the relationships between crop yields, water use, climate patterns, and economic pressures.

Mathematical techniques such as linear programming and optimization are used to allocate limited resources like land, water, and fertilizers. The goal is to support crop production while limiting environmental harm. Statistical modeling and machine learning are also used to predict yields, detect disease, and guide irrigation decisions under uncertain conditions.

The mathematics of food also includes the study of food supply chains. Mathematical models are applied to logistics, waste reduction, and food safety across long and often fragile distribution networks. Inventory management, transportation planning, and demand forecasting all rely on quantitative analysis to keep food moving reliably.

In food processing, mathematical methods are used to support both safety and quality. Models help determine production conditions, estimate shelf life, and analyze processes such as sterilization and pasteurization. Computational simulations allow these processes to be tested without risking contamination or product loss.

Mathematics also plays an important role in nutrition. Meal planning and dietary optimization often rely on mathematical techniques to balance nutritional needs with cost, availability, and environmental impact. Linear programming and related methods are used to support both individual health goals and broader public nutrition efforts.

The mathematics of food intersects with economics as well. Mathematical models are used to study food markets, anticipate price changes, and analyze the effects of policy decisions. Game theory helps explain how producers, processors, retailers, and consumers interact, often in ways that shape access and affordability.

Climate change and sustainability are central concerns within the mathematics of food. Models are used to examine how rising temperatures, shifting precipitation, and extreme weather events affect food systems. These tools also support the development of adaptation strategies and climate-resilient agricultural practices.

The growth of precision agriculture has increased the use of mathematical techniques such as GPS and GIS in crop management. Data analytics and machine learning are used to interpret large datasets related to soil health, weather patterns, and yield variability. This allows decisions to be based on observed conditions rather than averages.

Ultimately, the mathematics of food is concerned with outcomes that extend beyond numbers. Its purpose is to support food systems that are sustainable, equitable, and capable of meeting human needs. Mathematical tools help address complex challenges while keeping long-term impacts in view.

The mathematics of food is especially relevant today because global food systems face increasing pressure. With the world's population projected to reach 9.7 billion by 2050, demand for food is expected to rise sharply. Meeting this demand requires not only increased production, but better use of resources.

One reason mathematics is essential in food systems is its role in **resource allocation**. Mathematical models help farmers and policymakers decide how land, water, and fertilizers can be used more efficiently, reducing waste and environmental damage. Precision agriculture, which relies heavily on algorithms, is one example of this approach.

Mathematics is also critical for understanding and responding to **climate change**. As temperatures rise and weather patterns shift, food production becomes more uncertain. Mathematical models help anticipate these impacts and guide adaptation strategies before losses occur.

Food security and nutrition are another major area where mathematics matters. Models can support nutrition planning that accounts for dietary needs, cost, and availability. These approaches are used to address both malnutrition and diet-related health conditions affecting populations worldwide.

The mathematics of food also has implications for **economic development**, particularly in regions where agriculture remains central to livelihoods. Mathematical analysis supports decisions about investment, trade, pricing, and market regulation.

Reducing food waste is another significant challenge. A large share of global food production is lost before it reaches consumers. Mathematical models help identify inefficiencies in supply chains and support strategies that improve storage, transportation, and distribution.

Mathematics also helps explain how food systems operate as **interconnected networks**. Producers, processors, distributors, retailers, and consumers influence one another in complex ways. Modeling these interactions supports more informed policy decisions and market interventions.

Because of this complexity, the mathematics of food depends on **collaboration across disciplines**. Mathematicians, agronomists, economists, policymakers, and industry professionals all contribute different forms of expertise. When these perspectives are aligned, mathematical tools become practical rather than theoretical.

The mathematics of food is therefore central to humanity's future. By applying mathematical reasoning to real constraints and real systems, it offers ways to build food systems that are more resilient, more equitable, and better able to meet global needs.

Designing a Mathematical Equation for Food Journey

Let's create a simplified equation representing the food journey from production to digestion. We'll call it the Food Journey Equation (FJE):

$$FJE = (P \times Q \times N) / (T \times W \times D)$$

Where:

- P = Production efficiency (yield, sustainability, etc.)
- Q = Quality (nutritional value, taste, etc.)
- N = Nourishment potential (digestibility, bioavailability, etc.)
- T = Time (from production to consumption)
- W = Waste (loss, spoilage, etc.)
- D = Digestion efficiency (individual factors, gut health, etc.)

This equation is a basic framework. We can refine it further.

Modifying the Food Journey Equation (FJE)

Let's tweak the components:

- P = Production Sustainability (environmental impact, resource use)
- Q = Nutritional Quality (macro-nutrients, micro-nutrients, etc.)
- N = Nourishment Value (satiety, health impact, etc.)
- T = Time-to-Plate (freshness, processing, etc.)
- W = Waste Minimization (reduction, repurposing, etc.)
- D = Digestion Efficiency (gut health, individual factors)

Updated equation:

$$FJE = (P \times Q \times N) / (T \times W \times D)$$

We can add or adjust components (e.g., add "S = Sustainability" or "E = Environmental footprint")

Let's tweak the components:

- P = Production Sustainability (environmental impact, resource use)
- Q = Nutritional Quality (macro-nutrients, micro-nutrients, etc.)
- N = Nourishment Value (satiety, health impact, etc.)
- T = Time-to-Plate (freshness, processing, etc.)
- W = Waste Minimization (reduction, repurposing, etc.)
- D = Digestion Efficiency (gut health, individual factors)

Updated equation:

FJE = (P × Q × N) / (T × W × D)

We can also add or adjust components.

Let's add:

- S = Sensory Appeal (taste, texture, aroma)
- A = Accessibility (cost, availability, cultural relevance)

Updated equation:

FJE = (P × Q × N × S × A) / (T × W × D)

This adds more dimensions to the food journey.

Now let us prioritize components or explore applications (e.g., sustainable food systems, personalized nutrition)

Prioritizing Components & Applications

Let's prioritize components:

1. N = Nourishment Value (core focus on health impact)
2. Q = Nutritional Quality (nutrient density, etc.)
3. S = Sensory Appeal (influences consumption choices)
4. P = Production Sustainability (environmental impact)
5. A = Accessibility (equitable access to food)
6. D = Digestion Efficiency (individual factors)

7. T = Time-to-Plate (freshness, processing)
8. W = Waste Minimization (reduction, repurposing)

Applications:

- Sustainable Food Systems: Use FJE to optimize P, Q, N, and W for eco-friendly, nutritious food.
- Personalized Nutrition: Tailor A, D, and S for individual health needs.

Specify Applications (e.g., AgriConnect Africa projects)

Exploring AgriConnect Africa Applications

Let's explore how FJE applies to AgriConnect Africa:

- Optimize Production Sustainability (P): Promote regenerative agriculture practices among small-scale farmers.
- Enhance Nutritional Quality (Q): Focus on nutrient-dense crops, like biofortified varieties.
- Boost Nourishment Value (N): Educate on balanced diets and healthy eating habits.
- Improve Accessibility (A): Connect farmers to markets, reduce supply chain inefficiencies.

Let's prioritize specific areas or explore partnerships (e.g., with Universities Innovation Centre)

Potential Partnerships & Priorities with University Innovation Centres

- Research & Development: Collaborate on optimizing FJE components for AgriConnect Africa.
- Training & Capacity Building: Develop programs for entrepreneurs, farmers, and students.
- Innovation Incubation: Support AgriConnect Africa's digital platform and services.

Priorities:

1. Connect farmers to markets: Enhance A (Accessibility) and reduce W (Waste).
2. Promote sustainable practices: Boost P (Production Sustainability).
3. Improve nutrition outcomes: Focus on Q (Nutritional Quality) and N (Nourishment Value).

Agriconnect Africa

Prioritize Components & Applications with Symbols And Components

1. Nourishment Value (N) - α
2. Nutritional Quality (Q) - β

3. Sensory Appeal (S) - γ
4. Production Sustainability (P) - δ
5. Accessibility (A) - ε
6. Digestion Efficiency (D) - ζ
7. Time-to-Plate (T) - η
8. Waste Minimization (W) - θ

Exploring AgriConnect Africa Applications

$\alpha + \beta$ = Nourishment Value + Nutritional Quality

- Optimize Production Sustainability (P δ)
- Enhance Nutritional Quality (Q β)
- Boost Nourishment Value (N α)
- Improve Accessibility (A ε)

Potential Relationship with Universities

$\delta + \varepsilon$ = Production Sustainability + Accessibility

- Research & Development (R λ)
- Training & Capacity Building (T μ)
- Innovation Incubation (I ν)

Chapter 1: The Food Equation

The Poisoned Plate

In a world where food was cheaper and more abundant than ever, the tables of millions were filled with meals that promised convenience but delivered danger. Processed foods, pumped with additives and preservatives, became the norm. Then the reports started rolling in. Skyrocketing rates of obesity, diabetes, and heart disease followed. The food industry's focus on maximizing profit created a toxic cycle. Cheap food fueled illness, which fueled demand for more cheap food.

A mathematician, working with public health experts, began unraveling the numbers behind the crisis. They mapped how production shortcuts amplified risk, how supply chains spread contaminants, and how pricing strategies pushed people toward unhealthy choices. The data was stark. The math driving food systems had ignored human health. But it also held the key to change. By recalculating priorities and factoring in nutrition, safety, and equity, new models emerged. Food became not just abundant, but nourishing.

This shift marked the beginning of a revolution in how we think about food. No longer just a commodity, food became a system where mathematics could save lives.

The world's most pressing challenges—feeding a growing population, mitigating climate change, and ensuring sustainable development—are all intertwined with food. Food sits at the nexus where human health, environmental sustainability, and economic development meet. At the heart of this complex system lies mathematics.

Mathematics is the language of food. It describes the relationships between crops, climate, soil, water, and human societies. It helps explain the food equation, a complex interplay of factors that determine what we eat, how we produce it, and what it does to our bodies and the planet.

The Food Equation:

{Food} = {Crops, Climate, Soil, Water, Humans, Economics, Policy}

This equation represents the relationships between the factors that shape our food systems. Let's break it down:

- **Crops:** The varieties of plants cultivated for food, fiber, and fuel.

- **Climate:** The weather patterns and conditions that affect crop growth and yield.
- **Soil:** The foundation of agriculture, providing nutrients and support for crops.
- **Water:** The resource that sustains crops and agricultural systems.
- **Humans:** The farmers, producers, processors, distributors, and consumers who shape the system.
- **Economics:** The market forces that influence production, pricing, and trade.
- **Policy:** The regulations, laws, and agreements that govern food systems and agriculture.

The Math Behind Food

Mathematics is used to describe and analyze each component of the food equation. For example:

- **Crop modeling:** Statistical models predict crop yields, disease spread, and pest behavior.
- **Climate modeling:** Models forecast weather patterns, temperature shifts, and extreme events.
- **Soil science:** Mathematical models describe soil hydrology, nutrient cycling, and erosion.

- **Water management:** Optimization techniques allocate water resources, reducing waste and improving efficiency.

A Brief History of Food and Math

The connection between food and math dates back to ancient civilizations, when farmers used basic arithmetic to manage crops and trade goods. Today, advanced techniques such as machine learning, data analytics, and optimization are transforming food systems. These tools help produce more with less, reduce waste, and support sustainability.

The Power of Math in Food Systems

Mathematics has the power to:

- **Optimize crop yields:** Improve performance, reduce yield gaps, and increase food availability.
- **Reduce waste:** Save resources, lower emissions, and strengthen food security.
- **Promote sustainability:** Support climate-resilient agriculture, conserve water, and protect biodiversity.

As we navigate the complexity of the food equation, mathematics remains a guiding force. It illuminates pathways toward a more sustainable, equitable, and food-secure future.

Chapter 2: Counting Crops

The dawn of agriculture marked a pivotal moment in human history, and mathematics has played a crucial role in this transition. From ancient farmers counting seeds to modern agronomists analyzing satellite data, numbers have always been essential to growing food. In this chapter, we explore how mathematical concepts such as probability, statistics, and geometry help us understand and optimize crop production.

The Origins of Agriculture and the Role of Math in Early Farming Practices

Around 10,000 years ago, humans began domesticating plants and animals, leading to settled agriculture. Early farmers used basic math to manage their crops—counting seeds, measuring land, and tracking seasonal patterns. These practices allowed them to optimize yields and ensure food security.

The Fertile Crescent: Cradle of Agriculture

The Fertile Crescent, a region in the Middle East, is often referred to as the "cradle of agriculture." Ancient farmers here domesticated wild emmer wheat, barley, and legumes, laying the foundation for modern farming.

- **Counting Seeds:** Farmers counted seeds to ensure the right amount was planted, maximizing yields and minimizing waste.
- **Measuring Land:** Land measurements helped optimize crop rotation, irrigation, and soil fertility.
- **Tracking Seasons:** Farmers tracked seasonal patterns to determine the best times for planting, harvesting, and other critical tasks.

How Probability Helps Farmers Manage Risk and Uncertainty

Agriculture is inherently uncertain. Weather, pests, and diseases can dramatically affect crop yields. Probability theory enables farmers to assess risks and make informed decisions.

Applications of Probability in Agriculture:

- **Weather Forecasting:** Probability models help predict weather patterns, guiding decisions on planting, irrigation, and harvesting.
- **Pest Management:** Predicting pest outbreaks allows for targeted interventions, reducing crop losses.

- **Crop Insurance:** Probability informs crop insurance policies, helping farmers manage financial risk and protect livelihoods.

Statistical Models for Predicting Crop Yields and Detecting Disease

Statisticians use historical data and machine learning to predict crop yields and detect diseases, enabling better crop management, reduced losses, and improved food security.

Applications of Statistical Models:

- **Yield Prediction:** Statistical models estimate crop yields, informing decisions about harvesting, storage, and marketing.
- **Disease Detection:** Patterns identified through statistics allow for targeted interventions, reducing pesticide use.
- **Precision Agriculture:** Models optimize fertilizer application, irrigation, and pruning, improving crop quality and reducing waste.

Geometric Patterns in Plant Growth and Their Applications

Plants often display geometric patterns, such as the Fibonacci spiral in sunflowers. Understanding these patterns allows farmers to optimize planting densities and pruning strategies.

The Fibonacci Spiral:

The Fibonacci spiral appears in many plants, including sunflowers, daisies, and pineapples. This pattern allows for maximum packing efficiency, helping plants use space and resources effectively.

Applications in Agriculture:

- **Planting Densities:** Geometric patterns inform optimal planting densities, improving yields and reducing waste.
- **Pruning Strategies:** Geometric insights guide pruning, promoting healthy growth and maximizing production.

Challenges to the Adoption of Mathematics in Farming

Despite its potential, math has not always been widely adopted in agriculture due to several challenges:

- **Limited Access to Data:** Historically, reliable data on weather, soil, and crops was scarce.
- **Complexity of Agriculture:** Agriculture involves many interacting variables, making accurate models difficult.

- **Cost and Accessibility:** Advanced tools and expertise were often too expensive for small-scale farmers.
- **Traditional Practices:** Many farmers relied on intuition or established methods rather than mathematical models.

These challenges led to consequences such as:

- **Inefficient Resource Allocation:** Without data-driven insights, resource use was often suboptimal.
- **Poor Risk Management:** Reliance on tradition sometimes led to misjudged weather, pests, or market fluctuations.
- **Limited Scalability:** Farms struggled to scale efficiently without mathematical guidance, impacting profitability.

Banking and Risk Management in Agriculture

Banks financing farmers face significant risks due to the unpredictable nature of agriculture. Traditional lending approaches often struggled to account for weather-related losses, crop diseases, and market fluctuations.

To mitigate these risks, banks are exploring strategies such as:

- **Risk-Sharing Models:** Partnerships with governments, insurers, or other stakeholders to distribute risk.
- **Crop Insurance:** Tailored insurance products help reduce financial losses from unforeseen events.
- **Digital Platforms:** Technology streamlines lending, improves credit scoring, and enhances risk assessment.

Government initiatives also support these efforts:

- **Pradhan Mantri Fasal Bima Yojana (PMFBY):** Crop insurance covering natural disasters, pests, and diseases.
- **Kisan Credit Card (KCC):** Low-interest loans for crop production and related expenses.

By adopting these strategies, banks can better manage risks and increase lending to the agricultural sector, supporting farmers' growth, resilience, and sustainability.

In summary, mathematics has been integral to agriculture since its inception. From counting seeds to analyzing satellite data, math helps optimize crop production and ensure food security. While adoption has historically been limited by data access, complexity, cost, and tradition, modern technology and precision agriculture are transforming the landscape. Data-driven farming improves decision-making, resource allocation, risk management, and productivity, helping farmers build sustainable and resilient systems.

Chapter 3: Sowing the Seeds of Optimization

In agriculture, optimization is key to maximizing yields, minimizing waste, and promoting sustainability. Mathematical optimization techniques help farmers and policymakers make data-driven decisions, ensuring food security and environmental stewardship.

Linear Programming for Optimal Resource Allocation

Linear programming is a powerful tool for optimizing resource allocation in agriculture. For example, a farm in California's Central Valley might use linear programming to allocate land, water, and fertilizers among different crops (e.g., almonds, grapes, and tomatoes).

Example: A study by the University of California, Davis, applied linear programming to a 1,000-acre farm. The model recommended allocating 40% of the land to almonds, 30% to grapes, and 30% to tomatoes, resulting in a 15% increase in profits. This demonstrates how mathematical models translate multiple constraints into actionable strategies, balancing profit with resource availability.

Optimization of Crop Rotation and Intercropping Strategies

Crop rotation and intercropping are essential for maintaining soil health and reducing pests and diseases. Mathematical optimization helps farmers select rotations and intercropping patterns that balance productivity with ecological sustainability.

Example: Researchers at Iowa State University developed optimized crop rotation strategies for corn and soybeans. Their model recommended a 4-year rotation (corn-soybeans-corn-soybeans) with cover crops, reducing nitrogen fertilizer use by 20% and increasing yields by 10%. This approach shows how rotation planning can improve both soil quality and output efficiency.

Game Theory Applications in Agriculture

Game theory informs decision-making where resources are shared or limited. It helps farmers and policymakers determine optimal strategies for bidding, allocation, and negotiation.

Example: In Australia's Murray-Darling Basin, game theory was applied to allocate water rights among farmers. The model enabled more efficient allocation of a scarce resource by guiding bidding and negotiation processes, demonstrating how strategic frameworks can improve resource distribution.

Heuristics and Metaheuristics for Complex Problems

Heuristics and metaheuristics, such as genetic algorithms and simulated annealing, address complex optimization problems where traditional methods may be inadequate.

Example: At the University of Illinois, a genetic algorithm optimized irrigation scheduling for a corn farm. The model reduced water usage by 25% while maintaining yields, saving the farmer $50,000 per season. These approaches highlight how computational methods can handle multiple, interacting variables efficiently.

Case Study: The Israeli Kibbutz

Israeli kibbutzim have implemented optimization techniques across agricultural operations. By using linear programming and other methods, they have improved crop yields, reduced water consumption, and increased profitability. This case illustrates how systematic application of optimization methods can produce measurable operational benefits.

Precision Agriculture

Precision agriculture applies optimization to manage inputs like water and fertilizers at a fine scale, improving efficiency and reducing waste.

Example: A University of Nebraska–Lincoln study found that precision irrigation reduced water usage by 30% while maintaining yields. Such targeted interventions demonstrate the practical impact of optimization on resource efficiency.

Agricultural Supply Chain Optimization

Optimization also applies to logistics, storage, and inventory management, ensuring that products move efficiently from farm to market.

Example: A University of California, Berkeley, study used optimization algorithms to reduce transportation costs for a California strawberry producer by 20%. Efficient supply chain management complements farm-level productivity improvements, reducing waste and cost simultaneously.

Strawberry plantations going through auditing by the markets before shipment.

Infield tasting

Climate Change and Agriculture

Mathematical models predict the effects of climate change and guide mitigation strategies.

Example: Researchers at the International Institute for Applied Systems Analysis (IIASA) projected a 6% decrease in global wheat yields by 2050. Modeling such impacts allows policymakers and farmers to anticipate challenges and adjust practices proactively.

Agricultural Robotics and Automation

Optimization underpins autonomous systems in agriculture, including tractors and drones.

Example: The University of Illinois developed optimization algorithms for tractor navigation, reducing fuel consumption by 10%. Automated systems improve efficiency while minimizing environmental impact.

Food Loss and Waste Estimates

Food waste is a significant global problem:

- **Global average:** 20–30% of total food production is lost or wasted (≈1.3–2 billion metric tons annually).
- **Developing countries:** 10–40% lost during harvest, processing, and storage, valued at $310 billion annually.

- **Developed countries:** 5–10% wasted post-consumption, valued at $680 billion annually.

Per-hectare losses:

- **Cereals:** 10–20% lost, equivalent to 0.5–1 ton/ha.
- **Fruits and vegetables:** 20–40% lost, equivalent to 2–4 tons/ha.

Mathematical optimization can help reduce these losses by:

- Optimizing harvest timing: Reducing over- or under-harvesting.
- Improving storage and logistics: Minimizing spoilage and damage.
- Predictive analytics: Matching production to market demand.
- Precision agriculture: Adjusting inputs to reduce waste.

The Stakes Ahead

Global food demand is projected to increase by 60% by 2050, requiring a 70% increase in agricultural production. Even a 1% reduction in food waste could feed millions, lower greenhouse gas emissions, and conserve water and land. By optimizing crop yields, resource allocation, and supply chains, we can increase food availability, improve food security, and reduce environmental impact. The stakes are high, and action is urgent.

Chapter 4: The Numbers in Your Food

From farm to table, numbers silently shape almost every aspect of the food we consume. From the nutrients listed on a label to the environmental footprint of our meals, numerical data guide both our choices and the systems that produce the food. In this chapter, we explore the world of **food numeracy**, where mathematics meets nutrition, and reveal how understanding numbers can empower us to make healthier, more informed decisions.

Food Labels: Decoding the Numbers

Food labels are more than just words—they are a treasure trove of numerical information that reveals what we are truly consuming. Understanding these numbers helps us see beyond marketing claims and make choices that support our health. Key elements include:

- **Nutrition Facts Panel:** Lists serving size, calories, macronutrients (carbohydrates, protein, fat), and micronutrients (sodium, sugar, fiber).

- **% Daily Value (DV):** Indicates how much of each nutrient a serving contributes to the recommended daily intake.

- **Ingredients List:** Ingredients are listed in descending order by weight, with the most abundant ingredients first. This helps identify hidden sugars, salts, or fats.

By reading labels critically, we can compare products, spot unhealthy additives, and better manage our portion sizes.

Understanding Food Measurements

Food comes with numbers measured in various units, and understanding them is essential for both nutrition and cooking:

- **Weight:** Grams (g), milligrams (mg), ounces (oz)

- **Volume:** Milliliters (mL), liters (L), cups, teaspoons

- **Energy:** Calories (kcal), kilojoules (kJ)

Knowing these units allows us to interpret recipes accurately, manage serving sizes, and track nutritional intake efficiently.

The Math Behind Food Pricing

Numbers don't just measure nutrition—they also drive economics. Understanding how food prices are calculated helps us navigate budgets and make cost-effective choices:

- **Cost per Unit:** Comparing the price per gram, ounce, or serving helps identify the best value.
- **Markup:** Retail prices often include a markup on wholesale costs, sometimes doubling or tripling the original price.
- **Taxes and Subsidies:** Government policies influence prices, often affecting which foods are affordable and accessible.

Being numerically literate about food costs can prevent overspending and encourage smarter shopping decisions.

Food Safety: Numbers Matter

Food safety is another realm where numbers save lives. Every date, temperature, and microbial limit is based on scientific research:

- **Expiration Dates:** Best-by, sell-by, and use-by dates help prevent the consumption of spoiled foods.

- **Temperature Control:** Safe storage and handling temperatures minimize bacterial growth.

- **Microbial Limits:** Guidelines specify acceptable levels of bacteria, yeast, or mold to ensure food is safe to eat.

Ignoring these numbers can lead to serious illness, highlighting why food numeracy isn't just an academic skill—it's a survival skill.

The Environmental Impact of Food

Numbers also reveal the hidden costs of what we eat, from water use to greenhouse gas emissions:

- **Water Footprint:** Liters of water required per kilogram of food highlight the resource intensity of production.

- **Carbon Footprint:** Greenhouse gas emissions per kilogram show the climate impact of different foods.

- **Food Waste:** Percentage of food lost or discarded points to inefficiencies in our food system.

By understanding these figures, we can make choices that protect the planet while nourishing ourselves.

Making Informed Choices

When we understand the numbers behind food, we gain the power to make smarter, healthier, and more sustainable decisions:

- **Compare Products:** Choose options with better nutritional profiles and fewer harmful additives.
- **Manage Portion Sizes:** Control calorie intake, reduce waste, and prevent overconsumption.
- **Support Sustainable Systems:** Opt for environmentally friendly and locally sourced foods whenever possible.

The modern food system, however, often works against these goals. Corporate greed frequently prioritizes profit over nutrition, sustainability, and social responsibility. This has led to:

- **Unhealthy Products:** High-sugar, high-salt, and high-fat foods contribute to diet-related diseases.
- **Environmental Degradation:** Unsustainable practices harm ecosystems, pollute air and water, and waste valuable resources.

- **Exploitation:** Vulnerable communities often face unfair labor practices, limited access to healthy food, and food insecurity.

The consequences of profit-driven decisions are severe:

- **Public Health Crises:** Obesity, diabetes, and heart disease burden individuals and healthcare systems.

- **Environmental Disasters:** Climate change, deforestation, and water scarcity threaten global ecosystems.

- **Social Inequity:** Marginalized communities suffer disproportionately from unhealthy food environments and economic disparities.

A paradigm shift is necessary:

- **Prioritize People and Planet:** Focus on nutrition, sustainability, and ethical practices.

- **Transparency and Accountability:** Enforce labeling, corporate responsibility, and regulatory oversight.

- **Support Local, Sustainable Food Systems:** Encourage community-led initiatives and environmentally conscious practices.

It's time to reclaim our food system and put people and planet first.

Unhealthy Foods: The Hidden Dangers

Certain foods contribute heavily to rising diet-related diseases worldwide. Common culprits include:

- **Ultra-Processed Foods:** High in added sugars, saturated fats, and sodium, but low in essential nutrients. Examples: packaged snacks, frozen meals, sugary drinks.

- **Processed Meats:** Hot dogs, sausages, and bacon increase cardiovascular and cancer risks.

- **Sugary Drinks:** Sodas and sweetened beverages contribute to obesity, diabetes, and dental issues.

- **Refined Carbohydrates:** White bread, sugary cereals, and processed snacks can lead to weight gain and insulin resistance.

- **Foods High in Saturated and Trans Fats:** Found in fried and processed foods, these increase cholesterol and heart disease risk.

Some specific examples include fast food (burgers, fries, nuggets), packaged snacks (chips, cookies, candy), frozen meals (high sodium, preservatives), and sugary cereals.

The World Health Organization estimates:

- **600 million people** fall ill from contaminated food annually.
- **420,000 deaths** result from foodborne diseases each year.
- **$110 billion** is lost in productivity and medical costs due to unsafe food in low- and middle-income countries.

Practical Tips for Healthy Eating

To protect health and reduce risks from unhealthy foods:

- **Read Food Labels:** Identify added sugars, sodium, and unhealthy fats.
- **Choose Whole Foods:** Prioritize fruits, vegetables, whole grains, and lean proteins.

- **Cook at Home:** Preparing meals from fresh ingredients gives control over nutrition.

Being mindful about what we eat can prevent chronic diseases and improve overall well-being.

The Economic Cost of Unhealthy Diets

The financial impact of poor diets is staggering. Globally, diet-related chronic diseases—cardiovascular conditions, cancers, and diabetes—drain healthcare systems and productivity:

- **Healthcare Costs:** In the U.S., unhealthy diets cost approximately **$233.3 billion annually**, with projected cardiovascular disease costs reaching **$2 trillion by 2050**.

- **Productivity Losses:** Poor diets result in **$184.6 billion in lost productivity** each year.

- **Global Burden:** WHO estimates that **32 million deaths** annually are linked to ultra-processed food consumption.

- **UK Costs:** Obesity-related health issues cost the NHS over **£11.4 billion annually**, with wider societal costs estimated at **£74.3 billion**.

These figures underscore the urgent need for healthier choices, policy reforms, and public health interventions to reduce both economic and human costs of diet-related diseases.

Here are some more facts from the CDC;

Fast Facts: Health and Economic Costs of Chronic Conditions

Chronic diseases account for most illnesses, disability, and deaths in the United States and are the leading drivers of health care costs.

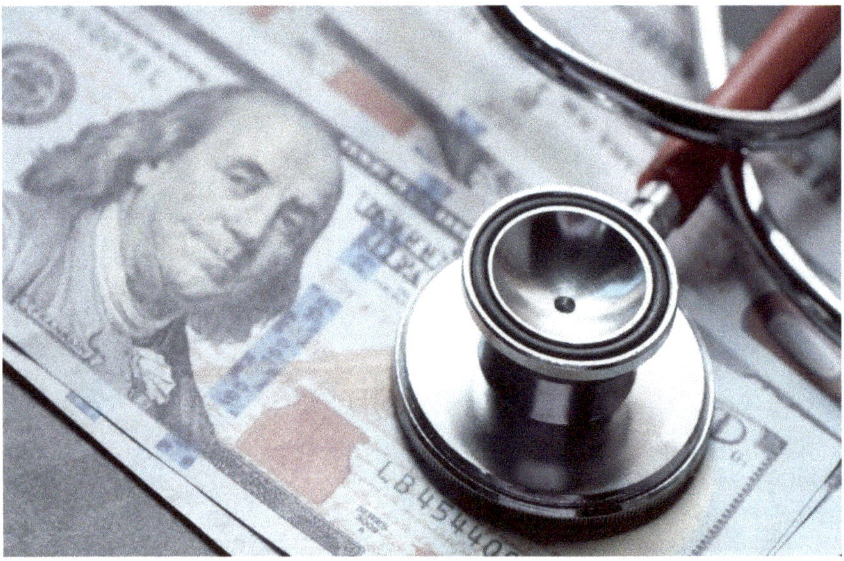

The Impact of Chronic Diseases in America

Chronic diseases dominate the health and economic landscape of the United States. **Ninety percent of the nation's $4.9 trillion in annual health care expenditures** are for people living with chronic or mental health conditions. Interventions to prevent and manage these diseases can yield significant benefits—both for individual health and for the economy.

- **Heart Disease and Stroke**

Nothing kills more Americans than **heart disease and stroke**. Over **843,000 Americans die** from these conditions each year—more than one in four deaths. Beyond the human toll, these diseases carry a massive economic burden: **$233.3 billion** in healthcare costs annually and **$184.6 billion** in lost productivity. By 2050, costs related to cardiovascular diseases are projected to hit roughly **$2 trillion**.

High blood pressure interventions, lifestyle modifications, and early detection programs demonstrate both health and economic benefits by reducing deaths and healthcare expenditures.

- **Cancer**

Each year, **1.8 million Americans** are diagnosed with cancer, and more than **600,000** die from it, making cancer the second leading cause of death. The cost of cancer care continues to climb, projected to exceed **$240 billion by 2030**. Early screening and targeted interventions for cancers such as breast, cervical, colorectal, and skin cancer can significantly reduce both mortality and financial burden.

- **Diabetes**

Over **38 million Americans** have diabetes, with another **98 million adults** classified as prediabetic, putting them at risk for type 2 diabetes. Diabetes can lead to serious complications, including heart disease, kidney failure, and blindness. In 2022, the total estimated cost of diagnosed diabetes—including medical care and lost productivity—was **$413 billion**. Effective interventions in diet, physical activity, and medical management can drastically reduce both human and economic costs.

- **Obesity**

Obesity affects **21% of children** and **40% of adults**, increasing the risk of type 2 diabetes, heart disease, and certain cancers. Alarmingly, only two in five young adults are physically prepared for military basic training. Obesity-related healthcare costs in the U.S. total nearly **$173 billion annually**, making prevention programs and lifestyle interventions critical.

- **Alzheimer's Disease**

Alzheimer's disease, a progressive and irreversible brain disorder, affects nearly **7 million Americans**, including one in nine adults aged 65 and older. Two-thirds of these older adults—about **4.2 million women**—are affected. Deaths from Alzheimer's more than doubled between 2000 and 2021, rising **141%**. The cost of caring for those with Alzheimer's and other dementias is expected to cost **$360 billion in 2024**, with projections nearing **$1 trillion by 2050**. This growing burden shows the urgent need for investment in research, early detection, and innovative care solutions to improve outcomes and reduce societal and economic impacts.

- **Arthritis**

Arthritis affects **53.2 million adults** in the United States—about one in five. It is a leading cause of disability and chronic pain. Medical costs and productivity losses attributable to arthritis exceeded **$300 billion in 2013**, highlighting the need for preventive care and management strategies.

- **Chronic Kidney Disease (CKD)**

CKD affects over **35.5 million adults**, with diabetes and hypertension as leading causes. Awareness is low, with nearly **9 out of 10 people unaware** they have CKD. Among those with severe CKD, **one in three** remains undiagnosed. Nearly **one in four Medicare dollars**, totaling **$95.7 billion**, is spent on CKD care.

- **Epilepsy**

Approximately **3 million adults** and **500,000 children and teens** under 18 have active epilepsy in the U.S. Adults with epilepsy report worse physical and mental health and more work limitations than those without. In 2019, total healthcare costs related to epilepsy reached **$13.4 billion**, with **$5.4 billion directly attributable** to the disease. Addressing these challenges through improved treatment, access to care, and support services could enhance quality of life and reduce the economic burden on families and the healthcare system.

- **Tooth Decay**

Cavities (tooth decay) are among the most common chronic diseases. **One in six children** aged 6–11 and **one in four adults** have untreated cavities. Untreated dental disease can lead to pain, infections, and impaired learning. Each year, **34**

million school hours are lost due to emergency dental care, and productivity losses approach **$46 billion**.

Key Risk Factors

Chronic diseases are often influenced by modifiable risk factors, which carry both health and financial consequences:

- **Cigarette Smoking:** Over **16 million Americans** have a disease caused by smoking, costing **$240 billion annually** in healthcare spending that could be prevented through cessation and prevention programs.

- **Physical Inactivity:** Inadequate activity contributes to heart disease, type 2 diabetes, obesity, and some cancers, costing **$192 billion annually**.

- **Excessive Alcohol Use:** Responsible for **178,000 deaths** annually, including one in five among adults aged 20–49. In 2010, excessive alcohol use cost the U.S. economy **$249 billion**, with three-quarters attributed to binge drinking.

This comprehensive look at chronic diseases underscores the **enormous health, social, and economic impact** they have on the nation. Addressing these diseases through prevention, lifestyle interventions, early detection, and equitable healthcare access is essential to improving both lives and the nation's economic well-being. Investing in comprehensive chronic disease strategies not only reduces long-term healthcare costs but also strengthens workforce productivity and societal resilience.

Chapter 5: The Maths Behind Meals

Numbers are everywhere in the food we eat, whether we notice them or not. From counting calories to calculating environmental impact, understanding the maths behind our meals can change the way we approach food, health, and sustainability. In this chapter, we will explore the numerical side of your plate, from ingredients and nutrition to production and environmental consequences, and how understanding these numbers can help you make smarter choices.

The Recipe for a Meal

To see the maths in action, let us consider a meal many of us enjoy: a cheeseburger with fries. By breaking it down into its components, we can examine the calories, protein, fat, and other nutritional elements of each ingredient.

Ingredients:

- 1 beef patty (100g)
- 1 bun (50g)
- 1 slice of cheese (20g)

- 1 serving of fries (100g)
- Ketchup and other condiments

Nutritional Information:

- Beef patty: 250 calories, 20g protein, 18g fat
- Bun: 120 calories, 4g protein, 2g fat
- Cheese: 100 calories, 6g protein, 7g fat
- Fries: 200 calories, 3g protein, 10g fat

Calculating the Total Nutritional Value

By adding these numbers, we can understand the total impact of the meal on our bodies:

- **Calories:** 250 + 120 + 100 + 200 = 670 calories
- **Protein:** 20g + 4g + 6g + 3g = 33g protein
- **Fat:** 18g + 2g + 7g + 10g = 37g fat

Even this simple calculation reveals something important, a meal is not just a source of energy, it is a balance of nutrients. Understanding these numbers helps us manage our diet better.

Nutrition is only part of the story. There is another layer of maths hiding behind every bite we take, the environmental cost of producing this food.

The Maths of Food Production

Every ingredient on your plate has a story that numbers can tell. Producing food requires water, land, energy, and creates greenhouse gases. Here is a closer look at the ingredients in our meal:

Beef Patty:

- 1 kilogram of beef requires roughly 1,000 liters of water to produce
- 1 kilogram of beef generates about 19 kilograms of CO_2 equivalent (CO_2e)

Fries:

- 1 kilogram of potatoes requires about 100 liters of water
- 1 kilogram of potatoes generates roughly 0.2 kilograms of CO_2e

Scaling this down to our meal:

Water usage:

100g beef patty, 100 liters of water

100g fries, 10 liters of water

Total: 110 liters

Carbon footprint:

100g beef patty, 1.9 kg CO2e

100g fries, 0.02 kg CO2e

Total: 1.92 kg CO2e

These numbers may seem abstract, but they represent real environmental consequences. Producing 1.9 kg of CO2e is roughly the same as driving a small car for 10 kilometers. A single meal carries a footprint that most of us rarely consider.

Why These Numbers Matter

Understanding the maths behind food is not just about curiosity; it directly affects our decisions:

- **Health:** Knowing calories, protein, and fat helps balance energy intake and supports overall wellbeing.
- **Environment:** Understanding water and carbon footprints allows us to reduce our ecological impact.

- **Food System Awareness:** Recognizing the hidden costs of production encourages more sustainable practices, both personally and on a larger scale.

By combining nutrition and environmental maths, we gain a complete picture of what eating a meal really entails.

Making Informed Choices

Numbers empower us to make smarter decisions. Here are some practical ways to apply this knowledge:

- **Choose sustainable options:** Plant-based proteins, seasonal vegetables, and locally sourced produce often have smaller environmental footprints.

- **Reduce waste:** Plan meals, store food correctly, and compost scraps whenever possible.

- **Prioritize nutrition:** Balance your calorie intake with nutrients. Opt for whole foods over processed options for better health outcomes.

- **Be mindful of portions:** Even healthy foods can have an environmental impact if consumed excessively.

For example, swapping a beef patty for a plant-based alternative could reduce water usage by up to 90 liters and lower carbon emissions by nearly 2 kilograms of CO2e per meal. Small changes add up over time.

Creating a Culture of Food Literacy

Understanding your meal's maths is not just a personal responsibility, it is a societal opportunity. Building awareness requires coordinated efforts:

1. **Food Education:** Schools can integrate food literacy into curricula, teaching children about nutrition, sustainability, and basic maths to quantify meals.

2. **Transparent Labeling:** Clear nutritional and environmental labels help consumers make informed choices.

3. **Public Awareness Campaigns:** Media, social platforms, and community programs can highlight the real impact of everyday meals.

4. **Industry Responsibility:** Companies should disclose environmental data and prioritize sustainable production.

5. **Digital Tools and Apps:** Interactive tools allow users to track nutrition, footprint, and food costs in real time.

6. **Policy and Regulation:** Governments can support food literacy, sustainability, and public health through legislation and incentives.

Key stakeholders include:

- Governments for policy and enforcement
- Food industry for transparency and sustainable production
- Educators for curriculum development
- Consumers for conscious decision-making

Together, these efforts help create a culture where meals are understood not just as fuel, but as choices that shape our health, communities, and planet.

And now we all know that the maths behind your meal is complex, but breaking it down reveals the true impact of what we eat. Each bite carries nutritional value, environmental cost, and social consequences. By understanding these numbers, we gain the power to make smarter choices that benefit our bodies, the planet, and the broader food system.

Food is more than taste; it is numbers in action. Learning to read these numbers and act on them is the first step toward a healthier, more sustainable, and informed relationship with what we eat.

Chapter 6: Food Systems and Networks

Understanding Networks, Challenges, and Solutions

Food systems encompass the entire process of food production, processing, distribution, and consumption. Understanding these complex networks is crucial for creating sustainable, equitable, and healthy food systems.

Food System Components

1. *Production*: Farming, livestock, and fisheries, including emerging techniques such as aquaponics and controlled-environment agriculture.

2. *Processing*: Transforming raw ingredients into food products, with value addition through fortification, branding, or packaging.

3. *Distribution*: Transportation, storage, and logistics to ensure timely delivery and minimal spoilage.

4. *Consumption*: Retail, food service, and household consumption shaped by cultural, economic, and dietary factors.

5. *Waste Management*: Food waste disposal and recovery, increasingly integrated with circular economy principles.

Food Networks

1. *Global Food Trade*: International trade agreements, tariffs, and logistics influence food availability and pricing.

2. *Supply Chains*: Relationships between producers, processors, distributors, and retailers determine efficiency and resilience.

3. *Local Food Systems*: Community-supported agriculture (CSA), farmers' markets, and local food hubs support regional self-reliance.

4. *Food Value Chains*: Adding value through processing, branding, and marketing enhances both economic opportunity and consumer choice.

Key Players

1. *Farmers and Producers*: Growers, ranchers, and fishermen.

2. *Food Processors and Manufacturers*: Companies transforming raw ingredients into food products.

3. *Distributors and Wholesalers*: Entities managing logistics, storage, and transportation.

4. *Retailers and Food Service Providers*: Supermarkets, restaurants, and institutional food services.

5. *Consumers*: Individuals and households purchasing and consuming food.

Challenges and Opportunities

1. *Sustainability*: Environmental impact, climate change, and resource depletion require careful management and ongoing innovation.

2. *Food Security*: Access, availability, and affordability of nutritious food remain key global concerns.

3. *Food Safety*: Ensuring safe handling, processing, and storage practices protects public health.

4. *Food Waste*: Reducing waste and recovering surplus food can improve efficiency and equity.

5. *Equity and Justice*: Promoting fair labor practices, equitable access, and cultural sensitivity helps ensure that food systems serve all populations.

Innovative Solutions

1. *Vertical Farming and Urban Agriculture*: Controlled-environment agriculture supports urban food production and reduces transport emissions.

2. *Precision Agriculture*: Technology-driven farming practices optimize yields and reduce inputs.

3. *Food Tech and E-commerce*: Digital platforms are transforming production, distribution, and retail.

4. *Circular Economy Models*: Closing loops and reducing waste increases sustainability.

5. *Collaborative Governance*: Multi-stakeholder partnerships and policy initiatives support integrated food system management.

Case Studies

1. *The Netherlands' Sustainable Food System*: National strategies promote sustainable agriculture, reduce waste, and improve public health.

2. *The Oakland Food Policy Council*: Community-led initiatives address food insecurity and strengthen local food systems.

3. *The Global Food Safety Initiative*: Multi-stakeholder partnerships improve food safety standards and practices.

Understanding food systems and networks enables us to identify opportunities for innovation, collaboration, and transformation, helping to create a more sustainable, equitable, and healthy food system.

Current Challenges

Environmental Challenges

- *Climate Change*: Rising temperatures, changing precipitation patterns, and extreme weather events impact agricultural productivity and food security.

- *Resource Scarcity*: Water scarcity, soil degradation, and biodiversity loss threaten long-term sustainability.

- *Pollution*: Agricultural practices contribute to greenhouse gas emissions, water pollution, and soil contamination[12].

Social and Economic Challenges

- *Food Insecurity*: Hunger and malnutrition affect over 800 million people globally.
- *Poverty and Inequality*: Small-scale farmers and vulnerable populations are disproportionately affected.
- *Food Waste*: Up to 30% of food produced globally is lost or wasted, worsening environmental and social challenges.

Health and Nutrition Challenges

- *Malnutrition*: Undernutrition, overnutrition, and micronutrient deficiencies are prevalent worldwide.
- *Diet-Related Diseases*: Poor diets contribute to rising obesity, diabetes, and other non-communicable diseases[13].

Policy and Governance Challenges

- *Lack of Coordination*: Multiple stakeholders and sectors complicate governance.

- *Inadequate Policies*: Short-term priorities often overshadow long-term sustainability and food security [45].

These challenges highlight the need for transformative approaches prioritizing sustainability, equity, and resilience.

Strategies for Overcoming Challenges

Sustainable Agriculture Practices

- *Agroecology*: Regenerative farming practices prioritize soil health, biodiversity, and ecosystem services.

- *Climate-Resilient Agriculture*: Promote drought-tolerant crops, conservation agriculture, and climate-smart techniques.

- *Reduce Chemical Use*: Encourage integrated pest management and organic methods [1][2].

Food Distribution and Access

- *Strengthen Local Food Systems*: Support small-scale farmers, local markets, and CSA programs.

- *Improve Food Infrastructure*: Invest in storage, transport, and logistics to reduce waste and increase access.

- *Food Assistance Programs*: Expand food stamps, school meals, and food banks [3][4].

Reducing Food Waste

- *Food Recovery*: Encourage donation and recovery programs.
- *Food Waste Reduction*: Implement technologies and practices to minimize loss in processing and retail.
- *Consumer Education*: Promote awareness and behavior change to reduce household waste [5].

Policy and Governance

- *Food Policy Councils*: Multi-stakeholder platforms enhance coordination and planning.
- *Sustainable Food Procurement*: Policies prioritize environmentally and socially responsible sourcing.
- *Food System Planning*: Integrate food system considerations into urban and regional development [6].

Supporting Small-Scale Farmers

- *Research and Extension*: Invest in programs benefiting small-scale farmers.

- *Market Access*: Improve value chains and infrastructure.
- *Farmer Support Programs*: Crop insurance, credit access, and cooperative development strengthen resilience.

Chapter 7: The Economies of Food

The economies of food are complex and multifaceted, influencing how we produce, distribute, and consume food. In this chapter, we'll explore the economic aspects of food systems, including market trends, trade policies, and the effects of globalization.

Food Markets and Trade

1. *Global Food Trade*: International trade agreements, tariffs, and logistics shape the global food market and determine how quickly and cheaply products move between regions.

2. *Food Prices*: Market fluctuations, supply and demand dynamics, and government policies drive food prices and affect affordability for consumers.

3. *Commodity Markets*: Global markets for staple crops such as wheat, corn, and soybeans play a major role in shaping food prices, availability, and producer decision-making.

Food Industry Structure

1. *Agribusiness*: Large-scale agricultural companies dominate production, processing, and much of the input supply chain, shaping market structure and bargaining power.

2. *Food Processing and Manufacturing*: Firms transform raw ingredients into consumer food products, adding value but also concentrating portions of the value chain.

3. *Retail and Food Service*: Supermarkets, restaurants, and other food establishments are the primary points of sale to consumers and play a major role in shaping demand, pricing, and food accessibility.

Economic Impact of Food Systems

1. *Employment*: Food systems provide employment across agriculture, processing, distribution, retail, and services, often forming the backbone of rural labour markets.

2. *GDP Contribution*: Food systems contribute significantly to national economies through production, trade, and value-added activities.

3. *Environmental and Health Costs*: Food systems impose external costs—environmental degradation, resource depletion, and healthcare expenses—that are frequently not reflected in market prices.

Challenges and Opportunities

1. *Market Volatility*: Price volatility in food markets can threaten food security and livelihoods, particularly for low-income households and smallholder farmers.

2. *Trade Tensions*: Protectionism, trade disputes, and sudden policy shifts can disrupt food trade and raise costs for importing countries.

3. *Sustainability and Equity*: Food systems must balance economic growth with social equity and environmental sustainability to be resilient and inclusive.

Case Studies

1. *The US Farm Bill*: A comprehensive agricultural policy that shapes production incentives, support programs, and trade impacts.

2. *The EU's Common Agricultural Policy*: A regional framework promoting sustainable agriculture, rural development, and market stability.

3. *Food Hubs and Cooperatives*: Community-led initiatives that support local food systems, improve market access for small producers, and promote fairer pricing.

By understanding the economies of food, we can identify opportunities for sustainable development, equitable trade, and healthier food systems. This understanding also enables policymakers, producers, and consumers to make informed decisions that strengthen food security while minimizing environmental and economic risks.

The global food balance is a complex issue, influenced by factors such as population growth, income changes, dietary preferences, and climate change. Below are some key trends.

Global Food Trends

- *Increasing Food Production*: Global food production has increased by 61% since 1990, outpacing population growth of 45%.

- *Shifting Diets*: As incomes rise, diets typically shift toward more animal products, fruits, and vegetables, changing demand patterns and supply requirements.
- *Food Waste*: Up to 30% of food produced globally is lost or wasted across production, handling, and consumption stages.

Regional Variations

- *Asia and Africa*: Experiencing rapid growth in both food production and consumption as agricultural investment and demand rise.
- *Europe and North America*: Observing a plateau or modest decline in per capita food supply in some categories, with a shifting focus toward quality and sustainability.
- *Sub-Saharan Africa:* Faces substantial food insecurity and malnutrition challenges, alongside opportunities for agricultural transformation. [1] [2]

Challenges Ahead

- *Population Growth*: The global population is projected to reach 9.7 billion by 2050, increasing demand for food and putting pressure on production systems.
- *Climate Change*: Changing weather patterns and extreme events threaten agricultural productivity and local food security.
- *Inequality*: Unequal access to nutritious food persists within and between countries, exacerbating health and economic disparities.

Opportunities for Improvement

- *Sustainable Agriculture*: Promoting practices that protect ecosystems, improve soil health, and raise long-term productivity.
- *Reducing Food Waste*: Strengthening supply chains, improving storage and transport, and addressing consumer-level waste can recover significant quantities of food.

- *Supporting Small-Scale Farmers*: Enhancing access to markets, finance, and technology improves food security and livelihoods. [3] [4] [5]

The global food balance requires a multifaceted approach that addresses production, distribution, and consumption simultaneously.

The economics of GMO foods are similarly complex and multifaceted. The global genetically modified food market is projected to reach $314.3 billion by 2035, growing at a compound annual growth rate (CAGR) of 7.9% from 2026 to 2035. This growth is driven by increasing demand for higher-yielding, resilient crops and ongoing advances in biotechnology.

Key Drivers:

- *Food Security*: GMO crops can help address food security by increasing yields and improving resistance to pests, diseases, and environmental stress.

- *Sustainability*: Certain GMO varieties can reduce pesticide and fertilizer needs, potentially lowering environmental impacts when used as part of integrated practices.

- *Economic Benefits*: Higher yields and lower input costs can increase farmer incomes and contribute to lower consumer prices when benefits are widely adopted.

Regional Trends:

- *North America*: Dominates GMO crop production, with the United States and Canada among the largest producers and adopters.

- *Asia-Pacific*: Emerging as a significant market as countries increasingly experiment with or adopt biotech crops to boost productivity.

- *Europe*: Adopts a more cautious regulatory stance toward GMOs, leading to slower uptake and tighter controls in many countries.

Future Outlook:

- *Increased Adoption*: Wider adoption of biotech and gene-edited crops is likely where regulatory frameworks and market acceptance allow.

- *Advancements in Biotechnology*: Gene-editing tools such as CRISPR are improving yield potential, stress tolerance, and nutritional profiles.
- *Regulatory Frameworks*: Policymakers will need to weigh safety assessments, trade implications, and public acceptance when shaping future rules. [1] [2] [3]

Overall, the economics of GMO foods appear promising in terms of productivity gains and innovation, but regulatory decisions and consumer acceptance will crucially shape the industry's trajectory.

The future of food economics in Sub-Saharan Africa shows significant potential for growth, with agricultural productivity projected to rise by around 3.2% annually over the next five years in optimistic scenarios. When combined with increased value-added processing and export development, agriculture could become a stronger engine of economic growth and rural development. [1]

Key Trends Shaping the Future:

- *Population Growth*: Rapid demographic expansion will raise food demand; Africa's population is expected to grow substantially by mid-century.

- *Urbanization*: Rising urban populations will increase demand for higher-value, processed, and convenient foods.

- *Climate Change*: Variability and long-term shifts in climate will affect yields, requiring adaptation and resilience-building.

- *Economic Diversification*: Diversifying economies and improving productivity in agriculture and agribusiness are central to sustained growth. [2] [3]

Opportunities and Challenges:

- *Agricultural Investment*: Domestic private investment in food and agriculture often exceeds foreign private investment and development assistance, highlighting the importance of local capital flows.

- *Youth Employment*: The agri-food sector can create jobs for young people, helping to reduce unemployment and poverty if value chains are strengthened.

- *Food Systems Transformation*: Transforming systems to be more equitable, nutritious, environmentally sustainable, and resilient will be critical for rural well-being. [4] [1]

To realise this potential, Sub-Saharan Africa must tackle climate risks, infrastructure gaps, and inequality while capitalizing on opportunities in market integration, technology adoption, and value addition.

The impact of AI on food distribution economics is poised to be transformative, as AI systems optimise logistics, reduce waste, and improve supply chain efficiency. Below are key benefits and applications.

Key Benefits of AI in Food Distribution

- *Cost Reduction*: AI-driven automation and predictive analytics can lower inventory holding costs, reduce transportation expenses, and streamline warehouse operations.

- *Improved Efficiency*: Intelligent systems enhance demand forecasting, automate inventory decisions, and speed up order fulfilment for more reliable deliveries.

- *Enhanced Sustainability*: AI can help curb food waste through better timing and routing, and by identifying inefficiencies across the supply chain.

- *Increased Transparency*: Real-time tracking and analytics improve visibility, enabling faster responses to disruptions and better decision-making. [1] [2] [3]

Applications of AI in Food Distribution

- *Demand Forecasting*: Predictive models improve the accuracy of demand estimates, reducing overstock and stockouts.

- *Inventory Management*: Automated systems maintain optimal stock levels, trigger reorders, and allocate goods efficiently.

- *Route Optimization*: AI-enabled logistics solutions plan fuel- and time-efficient delivery routes, lowering emissions and costs. [1] [3] [4]

Overall, AI promises to revolutionize food distribution economics by reducing costs, raising efficiency, and supporting more sustainable supply chains.

Chapter 8: Sustainability and Food Security

As the global population is projected to reach 9.7 billion by 2050, ensuring food security while promoting sustainability has become an urgent and interconnected challenge. This chapter explores the relationship between sustainability and food security, highlighting the environmental, social, and economic dimensions that shape this complex issue.

Sustainability Challenges

1. *Environmental Degradation*: Agriculture is a major contributor to greenhouse gas emissions, deforestation, biodiversity loss, and water pollution, placing significant pressure on natural ecosystems.

2. *Resource Scarcity*: Rising demand for food, water, land, and energy threatens the long-term sustainability of food systems, particularly in resource-constrained regions.

3. *Climate Change*: Increasing temperatures, shifting precipitation patterns, and a higher frequency of extreme

weather events disrupt agricultural productivity and heighten risks to food security.

Food Security Concerns

1. *Hunger and Malnutrition*: More than 800 million people worldwide experience food insecurity, while many others suffer from micronutrient deficiencies and poor diet quality.

2. *Poverty and Inequality*: Food insecurity disproportionately affects vulnerable populations, including small-scale farmers, women, children, and marginalized communities.

3. *Food Waste*: Up to 30% of food produced globally is lost or wasted, intensifying food insecurity while contributing to environmental degradation.

Sustainable Food Systems

1. *Agroecology*: Adopting regenerative farming practices that prioritize soil health, biodiversity, and ecosystem services to enhance long-term productivity.

2. *Climate-Resilient Agriculture*: Promoting drought-tolerant crops, conservation agriculture, and climate-smart practices to strengthen resilience to climate shocks.

3. *Food Value Chains*: Developing sustainable value chains by reducing post-harvest losses, minimizing food waste, and promoting fair labor practices.

Solutions and Strategies

1. *Sustainable Agriculture Practices*: Expanding conservation agriculture, agroforestry, integrated pest management, and efficient water use.

2. *Food System Transformation*: Advancing sustainable food systems that reduce waste, improve nutrition outcomes, and support inclusive economic growth.

3. *Policy and Governance*: Designing and implementing policies that support sustainable agriculture, food security, and climate resilience at local, national, and global levels.

Case Studies

1. *The Green Revolution*: A broad initiative aimed at improving agricultural productivity and food security, particularly in developing regions.

2. *The Sustainable Agriculture Initiative*: A global platform encouraging sustainable farming practices and reducing environmental impacts across supply chains.

3. *Food Banks and Food Assistance Programs*: Innovative approaches that address food insecurity while reducing food waste and supporting vulnerable populations.

By addressing sustainability and food security challenges together, it is possible to build a more equitable, resilient, and sustainable food system for all.

Food banks and food assistance programs play a vital role in reducing food insecurity and improving nutrition outcomes. Below is an overview of their functions and impact.

Food Banks

- *Definition*: Food banks are non-profit organizations that collect, store, and distribute food to individuals and households in need.

- *Services*: They provide emergency food assistance, support nutrition programs, and promote food recovery from surplus sources.

- *Impact*: Food banks help alleviate hunger, improve dietary intake, and support vulnerable populations during times of need.

Food Assistance Programs

- *Types*: These include government-funded programs (such as SNAP and WFP), non-profit initiatives, and community-based projects.

- *Goals*: Programs aim to reduce food insecurity, promote nutrition, and enhance household economic stability.

- *Examples*: School meal programs, food stamps, food vouchers, and targeted nutrition interventions.

Benefits and Challenges

- *Benefits*: Food assistance programs improve food security, reduce poverty, and contribute to better health and well-being.

- *Challenges*: Common obstacles include limited funding, logistical constraints, and ensuring equitable access for all populations.

Innovative Approaches

- *Food Recovery*: Redistributing surplus food that would otherwise be wasted to those facing food insecurity.
- *Nutrition Education*: Offering guidance on healthy eating, food preparation, and meal planning.
- *Digital Platforms*: Using technology to improve coordination, streamline distribution, and expand outreach.

By strengthening food banks and food assistance programs, societies can more effectively address food insecurity, promote nutrition, and build resilient communities.

Chapter 9: The Future of Food—Trends and Challenges

As we look ahead, the future of food is being shaped by a convergence of powerful forces, including technological innovation, environmental pressures, demographic shifts, and changing consumer values. Food systems are no longer defined solely by production and consumption; they are increasingly influenced by sustainability goals, health concerns, digital transformation, and social equity. This chapter explores the major trends shaping the future of food, the challenges ahead, and the pathways by which food systems can evolve to become more resilient, equitable, and sustainable.

Trends Shaping the Future of Food

1. Sustainable Diets:

Demand for sustainable diets is growing as consumers become more aware of the environmental and health impacts of their food choices. Plant-based, flexitarian, and low-carbon diets are gaining popularity due to concerns about climate change, animal welfare, and chronic disease. Governments, health organizations, and food companies are increasingly promoting dietary shifts that emphasize fruits, vegetables, legumes, and whole grains while reducing reliance on resource-intensive animal products.

2. **Food Tech and Innovation:**

Technological innovation is transforming how food is grown, processed, and distributed. Precision agriculture uses data, sensors, and AI to optimize input use, increase yields, and reduce environmental impact. Vertical farming and controlled-environment agriculture enable year-round production in urban settings, reducing land use and transportation emissions. At the same time, innovation in food processing and preservation is extending shelf life and improving food safety.

3. **Digitalization and E-commerce:**

Digital technologies are reshaping food markets and consumer behavior. Online grocery platforms, meal kit services, and food delivery apps have expanded rapidly, offering convenience and wider choice. Digital tools also improve traceability, transparency, and coordination across supply chains, enabling producers and retailers to respond more quickly to changes in demand.

4. **Personalized Nutrition:**

Advances in genomics, wearable technology, and artificial intelligence are making personalized nutrition increasingly feasible. Diets tailored to individual health profiles, lifestyles, and genetic predispositions are expected to play a growing role in disease prevention and overall well-being. This trend reflects a broader shift away from one-size-fits-all dietary guidance toward more targeted, data-driven approaches.

Challenges Ahead

1. **Climate Change and Sustainability:**

Climate change remains one of the most serious threats to global food systems. Rising temperatures, shifting rainfall patterns, and extreme weather events reduce crop yields, disrupt supply chains, and increase price volatility. Reducing greenhouse gas emissions from agriculture while adapting to climate impacts is essential for long-term food security.

2. **Food Security and Nutrition:**

Despite global progress in food production, hunger and malnutrition persist in many regions. At the same time, overconsumption and diet-related diseases such as obesity and diabetes are increasing. Addressing food security requires not only producing enough food but also ensuring access to nutritious, affordable, and culturally appropriate diets.

3. **Food Waste and Loss:**

Significant quantities of food are lost at every stage of the supply chain, from production and storage to retail and consumption. Food waste represents a loss of resources, income, and environmental capital. Reducing waste is one of the most cost-effective ways to improve food availability and sustainability.

4. **Equity and Inclusion:**

Inequality remains a structural challenge within global food systems. Small-scale farmers, agricultural workers, women, and marginalized communities often face limited access to land, finance, technology, and markets. Ensuring fair labor practices, inclusive value chains, and equitable access to nutritious food is central to building just and resilient food systems.

Emerging Technologies

1. **Gene Editing:**

Gene-editing technologies such as CRISPR are accelerating crop improvement by enabling precise modifications to plant traits. These tools can enhance yield potential, disease resistance, drought tolerance, and nutritional quality, offering solutions to climate and food security challenges.

2. **Alternative Proteins:**

The development of plant-based, lab-grown, and insect-based proteins is reshaping the protein sector. These alternatives offer potential environmental benefits, including lower greenhouse gas emissions and reduced land and water use, while meeting growing global protein demand.

3. **Vertical Farming and Urban Agriculture:**

Urban agriculture and vertical farming systems allow food to be produced closer to consumers. These systems reduce transportation costs, enhance freshness, and improve resilience in densely populated areas, particularly where arable land is limited.

Future Food Systems

1. **Circular Food Systems:**

Circular models aim to minimize waste and maximize resource efficiency by reusing by-products, recycling nutrients, and closing material loops. Examples include composting food waste, converting waste into energy, and designing packaging that is reusable or biodegradable.

2. **Regenerative Agriculture:**

Regenerative practices focus on restoring soil health, increasing biodiversity, and enhancing ecosystem services. By improving soil structure and carbon sequestration, regenerative agriculture contributes to climate mitigation while supporting long-term productivity.

3. **Food Systems Resilience:**

Building resilient food systems involves strengthening the capacity to absorb shocks, adapt to change, and recover from disruptions. This includes diversifying production, improving storage and infrastructure, and strengthening local and regional food networks.

By understanding these trends, challenges, and emerging technologies, policymakers, producers, and consumers can work together to shape food systems that support environmental sustainability, social equity, and human health.

Food systems are likely to undergo significant transformation in the coming years, driven by innovation, environmental constraints, and evolving consumer expectations. These changes will influence how food is produced, consumed, and governed.

Changes in Food Production

- *Alternative Proteins*: As technology advances and costs decline, alternative proteins are expected to become more mainstream, complementing traditional livestock production.
- *Vertical Farming and Urban Agriculture*: Controlled-environment agriculture will continue to expand, particularly in cities, enhancing local food availability and reducing supply chain vulnerabilities.
- *Gene Editing*: Improved crop varieties will support higher yields, better nutrition, and increased resilience to climate stress.

Changes in Food Consumption

- *Sustainable Diets*: Consumer preferences will increasingly favor foods with lower environmental footprints and higher nutritional value.
- *Personalized Nutrition*: Data-driven dietary guidance will influence food choices, healthcare, and product development.
- *Food Waste Reduction*: Greater awareness and improved technologies will encourage households and businesses to reduce waste.

Changes in Food Systems

- *Circular Food Systems*: Waste reduction and resource efficiency will become central principles in food system design.
- *Digitalization and E-commerce*: Digital platforms will further integrate producers and consumers, improving efficiency and transparency.
- *Regenerative Agriculture*: Long-term ecosystem health will become a defining goal of agricultural development.

A combination of consumer demand, technological progress, policy interventions, and environmental necessity will drive these shifts. Ultimately, the future of food will depend on collective choices made today, shaping food systems that are capable of nourishing a growing population while safeguarding the planet.

Chapter 10: The Power of Math in Food Systems

Mathematics plays a vital role in understanding and optimizing food systems. From crop yields to supply chain management, math helps us turn complex realities into numbers we can analyze, compare, and improve. Whether it's deciding how much fertilizer to use, how to price a product, or how to reduce waste, maths is the quiet engine behind smarter, fairer, and more efficient food decisions.

Math in Food Production

- **Crop Modeling:**

Mathematical models predict crop yields, estimate how weather and soil conditions affect growth, and test "what if" scenarios before farmers take action. By using equations and simulations, farmers can optimize irrigation, fertilizer use, and planting times, thereby reducing waste, protecting resources, and stabilizing income.

- **Precision Agriculture:**

Math-driven technologies like GPS, remote sensing, and drones collect data on soil moisture, plant health, and nutrient levels. Algorithms then process this data to decide exactly where and how much to irrigate or apply inputs. This targeted use of resources improves farming efficiency, reduces costs, and minimizes environmental impact.

- **Food Processing:**

Mathematical optimization techniques improve production planning, machine scheduling, and inventory management in processing plants. For example, linear programming can help allocate limited resources (such as labor, time, or storage) to minimize cost or maximize output, while queuing models can reduce bottlenecks in processing lines.

Math in Food Distribution

- **Supply Chain Optimization:**

Math-driven models route trucks, ships, and storage in ways that reduce transportation costs, improve delivery times, and minimize spoilage. Techniques such as network optimization and route planning (e.g., solving the "shortest path" and "traveling salesman" problems) are used to deliver food to its destination as efficiently as possible.

- **Inventory Management:**

Mathematical techniques help retailers, warehouses, and distributors decide how much stock to order and when. Forecasting models predict demand, while inventory formulas (like economic order quantity and safety stock models) reduce spoilage, avoid stockouts, and ensure food availability even when demand is uncertain.

Food Safety:

Math-based risk assessment and predictive analytics identify where contamination is most likely to occur and the severity of the consequences. Statistical models analyze patterns in outbreaks, lab results, and inspection data, enabling authorities and companies to act early, recall products more quickly, and reduce contamination risks.

Math in Food Policy and Decision-Making

- **Food Security Analysis:**

Mathematical models estimate future food demand based on population growth, income changes, and climate trends. These models help governments and organizations plan for emergencies, decide on imports and reserves, and design programs that ensure food availability and affordability.

- **Economic Analysis:**

Math-driven economic models evaluate the impact of policies, trade agreements, taxes, and subsidies. By simulating different scenarios—such as changing farm subsidies or introducing a sugar tax—decision-makers can see how prices, consumption, production, and health outcomes might shift.

- **Sustainability Assessment:**

Mathematical tools assess the environmental impact of food systems by quantifying greenhouse gas emissions, water use, land use, and biodiversity loss. Life cycle assessment (LCA) models, for example, compare the environmental "footprint" of different foods, helping to inform more sustainable farming, processing, and consumption practices.

The power of math in food systems is undeniable. By applying mathematical techniques, we can optimize food production, reduce waste, design better policies, and ensure food security for a growing global population—while also protecting the planet.

Conclusion

The future of food is complex and multifaceted. As we've explored in this book, math plays a crucial role in understanding and optimizing food systems from seed to plate. By embracing math and data-driven approaches, we can create a more sustainable, equitable, and resilient food system for all.

The journey doesn't end here. The power of math in food systems is just beginning to unfold. As we continue to innovate, build better data systems, and apply mathematical techniques, we'll unlock new opportunities to improve food security, reduce waste, and promote sustainability and a healthy world population.

Why are we where we are today?

The story of how we ended up here is complex and multifaceted. One major factor is the shift in global food systems over the past century. With the rise of industrial agriculture and mass production, food became cheaper, more convenient, and widely available. However, this shift also led to a reliance on processed and high-calorie foods, contributing to the growing obesity and diet-related disease epidemic. The very systems designed to feed the world efficiently have, in many ways, undermined long-term health.

Another significant factor is the changing nature of work and lifestyle. As people moved from rural areas to cities, their daily lives became more sedentary, and they relied more heavily on processed and fast food. Time pressures, long commutes, and increased screen time have contributed to a decline in physical activity and an increase in unhealthy eating habits. Convenience has often replaced cooking skills and traditional food cultures that once emphasized fresh, whole foods.

The food industry's marketing and lobbying efforts have also played a significant role. Companies have aggressively promoted unhealthy products, often targeting vulnerable populations like children and low-income communities. Highly processed foods are marketed as fun, affordable, and convenient, while the long-term health costs are hidden. This has helped shape consumer preferences and influenced policy decisions, perpetuating the status quo.

Additionally, government policies and subsidies have often prioritized the production of commodity crops such as corn and soy, which are often used in processed foods, animal feed, and sugary drinks. This has made unhealthy ingredients cheaper and more widely available, while healthier options like fruits, vegetables, and pulses are often more expensive and harder to access, especially in food deserts and low-income areas.

The result is a food environment that is stacked against healthy choices. Many people lack access to nutritious food, safe spaces for physical activity, and accurate information about what they eat. Misinformation, confusing labels, and aggressive marketing make it even harder to navigate. This has led to a global health crisis, with diet-related diseases like diabetes, heart disease, and certain cancers becoming increasingly prevalent.

So, how do we turn this around? It will take a collective effort from governments, industry leaders, scientists, and individuals to create a food system that prioritizes health, sustainability, and equity. This might involve policies like sugar taxes, marketing restrictions, and subsidies for healthy food producers. It could also mean investing in education and community programs that promote nutrition literacy, support local food systems, and rebuild cooking skills. Ultimately, it's about creating a food culture that values people and the planet over profits.

Approximately 420,000 people die every year due to foodborne diseases, with 30% of these deaths occurring among children under 5 years of age. This is a significant global health issue, with unsafe food causing around 600 million cases of foodborne diseases annually. In other words, nearly 1 in 10 people worldwide fall ill every year from contaminated food.

Regarding the percentage of the world population obsessed with food-related health issues, there isn't a precise global figure. However, it's clear that food safety and nutrition are major concerns globally, with many people affected by diet-related diseases like obesity and diabetes, while others struggle with undernutrition and micronutrient deficiencies. In many households, daily decisions about what to eat are directly tied to health, budget, and survival.

Some alarming statistics:

- *Obesity*: 1 in 8 people worldwide are living with obesity, with 2.5 billion adults overweight and 890 million living with obesity. This places enormous pressure on health systems and reduces the quality of life.

- *Foodborne illnesses*: 1.35 million cases of Salmonella infections occur annually in the US alone, resulting in 26,500 hospitalizations and 420 deaths. Many similar pathogens affect millions of people globally each year.

- *Diet-related deaths*: Poor diets are responsible for over 12 million avoidable deaths globally each year[123], including from heart disease, stroke, type 2 diabetes, and certain cancers.

These numbers highlight the importance of prioritizing food safety, nutrition, and healthy eating habits to mitigate these risks—and without mathematics, this mission is impossible. We need data to identify problems, models to test solutions, and numbers to hold systems accountable.

There are concerns about the safety and nutritional value of food, especially in times of crisis when people are desperate. In such situations, low-quality or unsafe food may be distributed or consumed simply because it is all that is available. Although laws, regulations, and standards are in place to ensure healthy, balanced foods, implementing and enforcing them can be challenging, especially in low-resource settings or during emergencies.

Global Food Safety Standards

- The Codex Alimentarius Commission sets international standards for food safety, covering topics like food additives, pesticide residues, contaminants, and labeling requirements. These standards provide a common reference for global trade and consumer protection.
- The World Health Organization (WHO) and the Food and Agriculture Organization (FAO) work together to develop guidelines for food safety and nutrition,

supporting countries in building strong regulatory systems and surveillance.

- Countries like the US, EU, and India have their own regulatory frameworks, such as the FDA, EFSA, and FSSAI, which oversee inspections, approvals, and enforcement within their territories.

Food Labeling and Tracking

- Many countries require food labeling that includes nutritional information, ingredients, allergen warnings, and sometimes origin. This helps consumers compare products and manage allergies or medical conditions.

- Some initiatives, like the Global Food Safety Initiative (GFSI), promote traceability and transparency throughout the food supply chain by setting benchmarks and encouraging best practices.

- However, more can be done to ensure that food is tracked from "field to fork" and meets standardized health checks. In many places, traceability breaks down at multiple points, making it difficult to identify the source of contamination or fraud.

Addressing Food Insecurity

- Organizations like the World Food Programme (WFP) and UNICEF provide emergency food assistance and address malnutrition in crisis situations, focusing on vulnerable groups such as children and pregnant women.

- Governments and NGOs are working to strengthen food systems and improve nutrition, especially for vulnerable populations, through school meal programs, social protection schemes, and support for smallholder farmers.

Standardizing labels to track food from field to fork is a crucial step towards ensuring transparency, safety, and nutrition in our food systems. By implementing a comprehensive labeling system, we can empower consumers to make informed choices about the food they eat, while also promoting accountability throughout the supply chain. This approach can help reduce foodborne illnesses, minimize waste, and support sustainable agriculture practices.

A standardized labeling system would require every stakeholder, from farmers to manufacturers, to provide detailed information about the food they produce and handle. This would create a digital trail, enabling real-time tracking and monitoring of food products, and facilitating swift recalls when necessary. By leveraging technologies such as blockchain, unique identifiers, and data analytics, we can create an immutable record of food provenance, handling, and storage, ensuring it is safe, fresh, and nutritious.

Moreover, standardized labels would be a game-changer for vulnerable populations, such as those affected by food insecurity, allergies, or diet-related diseases. By having access to clear, accurate information about the food they eat, individuals can make choices that align with their health needs, cultural preferences, and values. As we strive to build more resilient and equitable food systems, standardizing labels to track food from field to fork is a critical step towards a healthier, more sustainable future for all.

Let's take eggs and beef as examples, and then we'll do apples and oranges. We'll use rough estimates for simplicity. These are not exact market figures but illustrative numbers to show how maths can track costs, nutrition, and energy use across the food chain.

Eggs

Paragraph 1: Production

- 1 dozen eggs ≈ , 1.5 kg feed, 0.5 kg water, 1 kWh energy
- Feed cost: $0.50, Water cost: $0.01, Energy cost: $0.10
- Total production cost: $0.61/dozen eggs
- Vitamin benefit: 1 dozen eggs ≈ , 72 g protein, 180 mg cholesterol, 40% DV vitamin D

Paragraph 2: Processing

- Transportation to processing plant: 100 km, 0.1 L diesel
- Packaging: 1 dozen eggs, cardboard, $0.20
- Labor: $0.30
- Total processing cost: $0.50/dozen eggs
- Vitamin retention: 95% (handling and packaging)

Paragraph 3: Distribution

- Transportation to store: 500 km, 0.5 L diesel
- Storage: refrigeration, $0.20

- Total distribution cost: \$0.40/dozen eggs
- Vitamin retention: 90% (storage and handling)

Paragraph 4: Retail

- Store markup: 20%
- Final cost: \$2.50/dozen eggs
- Vitamin benefit per egg: 6 g protein, 30 mg cholesterol, 3.3% DV vitamin D

Paragraph 5: Consumption

- 2 eggs (100 g) ≈ 140 calories, 12 g protein
- Vitamin D: 6.6% DV, B12: 20% DV
- Cost per serving (2 eggs): ≈ \$0.42

Maths:

- Production: 1.5 kg feed + 0.5 kg water + 1 kWh energy ≈ \$0.61
- Processing: \$0.61 + \$0.50 = \$1.11
- Distribution: \$1.11 + \$0.40 = \$1.51

- Retail: $1.51 × 1.2 = $1.81 (wholesale base) + additional store costs = $2.50
- Cost per egg: $2.50 / 12 ≈ $0.21
- Cost per 2-egg serving: ≈ $0.42

This simple breakdown shows how small costs accumulate at each stage and how we can use maths to estimate both price and nutritional value per serving.

Beef

Paragraph 1: Production

- 1 kg beef ≈ 10 kg feed, 1,000 L water, 10 kWh energy
- Feed cost: $5, Water cost: $1, Energy cost: $1
- Total production cost: $7/kg beef
- Vitamin benefit: 1 kg beef ≈ 250 g protein, 200 mg B12, 50% DV zinc

Paragraph 2: Processing

- Slaughter, cutting, packaging: $3/kg beef
- Labor: $2

- Total processing cost: $5/kg beef
- Vitamin retention: 90% (handling and packaging)

Paragraph 3: Distribution

- Transportation to store: 1,000 km, 1 L diesel
- Storage: refrigeration, $1
- Total distribution cost: $2/kg beef
- Vitamin retention: 85% (storage and handling)

Paragraph 4: Retail

- Store markup: 30%
- Final cost: $20/kg beef
- Vitamin benefit per 100 g serving: 25 g protein, 20% DV B12, 15% DV zinc

Paragraph 5: Consumption

- 100 g beef ≈ 250 calories, 25 g protein
- Vitamin B12: 20% DV, zinc: 15% DV
- Cost per serving: $2

Maths:

- Production: 10 kg feed + 1,000 L water + 10 kWh energy ≈ $7
- Processing: $7 + $5 = $12
- Distribution: $12 + $2 = $14
- Retail: $14 × 1.3 = $18.20 (wholesale base) + additional store costs and margins ≈ $20
- Cost per 100 g serving: $20 / 10 = $2

Here, maths shows that beef is resource-intensive: more feed, more water, and a higher cost per unit of protein than many other foods.

Apples

Paragraph 1: Production

- 1 kg apples ≈ 0.1 kg fertilizer, 0.1 kg pesticides, 1 kWh energy
- Fertilizer cost: $0.05, Pesticides cost: $0.05, Energy cost: $0.10
- Total production cost: $0.20/kg apples

- Vitamin benefit: 1 kg apples ≈ 10% DV vitamin C, 5% DV fiber

Paragraph 2: Processing

- Washing, sorting, packaging: $0.30/kg apples
- Labor: $0.20
- Total processing cost: $0.50/kg apples
- Vitamin retention: 95% (handling and packaging)

Paragraph 3: Distribution

- Transportation to store: 500 km, 0.5 L diesel
- Storage: refrigeration, $0.20
- Total distribution cost: $0.40/kg apples
- Vitamin retention: 90% (storage and handling)

Paragraph 4: Retail

- Store markup: 30%
- Final cost: $2/kg apples

- Vitamin benefit per apple (150 g): 10% DV vitamin C, 5% DV fiber
- Cost per apple: \$0.30

Paragraph 5: Consumption

- 1 apple (150 g) ≈ 70 calories, 0.3 g protein
- Vitamin C: 10% DV, fiber: 5% DV
- Cost per serving: \$0.30

Maths:

- Production: 0.1 kg fertilizer + 0.1 kg pesticides + 1 kWh energy ≈ \$0.20
- Processing: \$0.20 + \$0.50 = \$0.70
- Distribution: \$0.70 + \$0.40 = \$1.10
- Retail: \$1.10 × 1.3 = \$1.43 (wholesale base) + additional store costs ≈ \$2
- Cost per apple: \$2 / 6–8 apples ≈ \$0.30

Oranges

Paragraph 1: Production

- 1 kg oranges ≈ 0.1 kg fertilizer, 0.1 kg pesticides, 1 kWh energy
- Fertilizer cost: $0.05, Pesticides cost: $0.05, Energy cost: $0.10
- Total production cost: $0.20/kg of oranges
- Vitamin benefit: 1 kg oranges ≈ 50% DV vitamin C, 10% DV fiber

Paragraph 2: Processing

- Washing, sorting, packaging: $0.30/kg oranges
- Labor: $0.20
- Total processing cost: $0.50/kg of oranges
- Vitamin retention: 95% (handling and packaging)

Paragraph 3: Distribution

- Transportation to store: 500 km, 0.5 L diesel
- Storage: refrigeration, $0.20
- Total distribution cost: $0.40/kg of oranges

- Vitamin retention: 90% (storage and handling)

Paragraph 4: Retail

- Store markup: 30%
- Final cost: $2.50/kg of oranges
- Vitamin benefit per orange (150 g): 25% DV vitamin C, 5% DV fiber
- Cost per orange: $0.40

Paragraph 5: Consumption

- 1 orange (150 g) ≈ 60 calories, 0.3 g protein
- Vitamin C: 25% DV, fiber: 5% DV
- Cost per serving: $0.40

Maths:

- Production: 0.1 kg fertilizer + 0.1 kg pesticides + 1 kWh energy ≈ $0.20
- Processing: $0.20 + $0.50 = $0.70
- Distribution: $0.70 + $0.40 = $1.10

- Retail: \$1.10 × 1.3 = \$1.43 (wholesale base) + additional store costs ≈ \$2.50
- Cost per orange: \$2.50 / 6–8 oranges ≈ \$0.40

Examples of eggs, beef, apples, and oranges illustrate how a standardized system can track food products from field to fork, providing insights into costs, nutrition, and environmental impacts. This transparency can empower consumers, farmers, and policymakers to make informed decisions about food production, distribution, and consumption, and to compare trade-offs between price, health, and sustainability.

Disadvantages

The current food system has several disadvantages, even for those who watch what they eat:

- **Lack of transparency:**

Limited access to information about food production, processing, and distribution makes it hard to make informed choices. People often don't know how their food was grown, what chemicals were used, or how workers were treated.

- **Inconsistent labeling:**

Food labels often lack clear, accurate information about nutritional content, ingredients, and environmental impact.

Different countries and brands use different systems, which can confuse consumers rather than guide them.

- **Industrialized food production:**

Prioritizes efficiency and profit over nutrition, sustainability, and animal welfare. This can lead to monocultures, the overuse of chemicals, and practices that prioritize yield over long-term soil and ecosystem health.

- **Environmental degradation:**

Contributes to climate change, water pollution, and soil degradation. Agriculture is a major user of water and land, and poorly managed systems can emit large amounts of greenhouse gases.

- **Health consequences:**

Diets high in processed and unhealthy foods contribute to rising rates of obesity, diabetes, and heart disease. Even people who try to eat "healthy" may struggle because of hidden sugars, salt, and fats in everyday products.

- **Food waste:**

An estimated 30–40% of food produced is wasted, contributing to environmental and economic issues. This waste happens at every stage—from farm losses and storage issues to retail rejection and household leftovers.

- **Inequitable access:**

Healthy food options are often inaccessible or unaffordable for marginalized communities. This deepens social inequalities and creates "double burdens" where undernutrition and obesity can exist in the same community—or even the same household.

These issues affect everyone, regardless of their eating habits. A more transparent and sustainable food system can benefit individuals, communities, and the environment. Maths and data are essential in measuring these problems, tracking progress, and designing better interventions.

There are concerns about practices in animal agriculture, where animal waste is repurposed as feed for other animals. This can pose health risks to humans and raise serious animal welfare concerns. It also reflects a mindset that treats animals as units of production rather than living beings.

Examples include:

- Using chicken litter (a mix of manure, feathers, and bedding) in cattle feed

Processing pig waste for fish feed

These practices prioritize production efficiency and cost-cutting over consumer safety and animal welfare, highlighting the need for stricter regulations, better enforcement, and much greater transparency in food production.

Cow manure is a valuable organic fertilizer rich in nutrients such as nitrogen, phosphorus, and potassium. In Sub-Saharan Africa, it's commonly used to improve soil fertility and promote sustainable agriculture. However, concerns arise about its "organic" status due to cattle vaccines, antibiotics, and potential contaminants.

Nutrient-Rich Benefits:

- *Improves Soil Structure*: Adds organic matter, enhancing water retention, aeration, and root development.
- *Boosts Crop Yields*: Provides essential nutrients for plant growth and can reduce reliance on synthetic fertilizers.
- *Supports Beneficial Microorganisms*: Fosters a healthy soil ecosystem, which can improve long-term soil health and resilience.

Potential Concerns:

- *Vaccine Residues*: Some vaccines or treatments may leave residues in manure, potentially affecting soil and plant health if not managed properly.

- *Pathogen Contamination*: Improper handling or application can spread diseases to humans, animals, or crops.

- *Heavy Metals*: Manure may contain heavy metals, such as copper and zinc from cattle feed supplements, which can accumulate in soils over time.

Best Practices:

- *Proper Handling and Storage*: Regularly collect, store, and apply manure to minimize nutrient loss, odors, and runoff.

- *Composting*: Break down manure through composting to kill pathogens, reduce contaminants, and stabilize nutrients.

- *Application Guidelines*: Follow local regulations and guidelines for safe application rates and timing to prevent soil overloading and water pollution.

In summary, cow manure can be a valuable organic fertilizer when managed properly. By understanding its benefits and potential concerns, farmers can harness its nutrient-rich properties while minimizing risks to health and the environment.

Some industries might benefit from the status quo, which could explain why they're hesitant to adopt transparent labeling or more rigorous traceability systems. By ignoring or downplaying labeling, they may avoid scrutiny, negative publicity, and potential market-share losses. This lack of transparency can make it harder for consumers to make informed choices and for policymakers to implement effective regulations.

Greed can hinder efforts to create a safer, more equitable world. When powerful entities prioritize profits over people and the planet, they undermine trust in institutions and perpetuate harm. This isn't just about individual choices—it is about systems that reward short-term gain and ignore long-term damage.

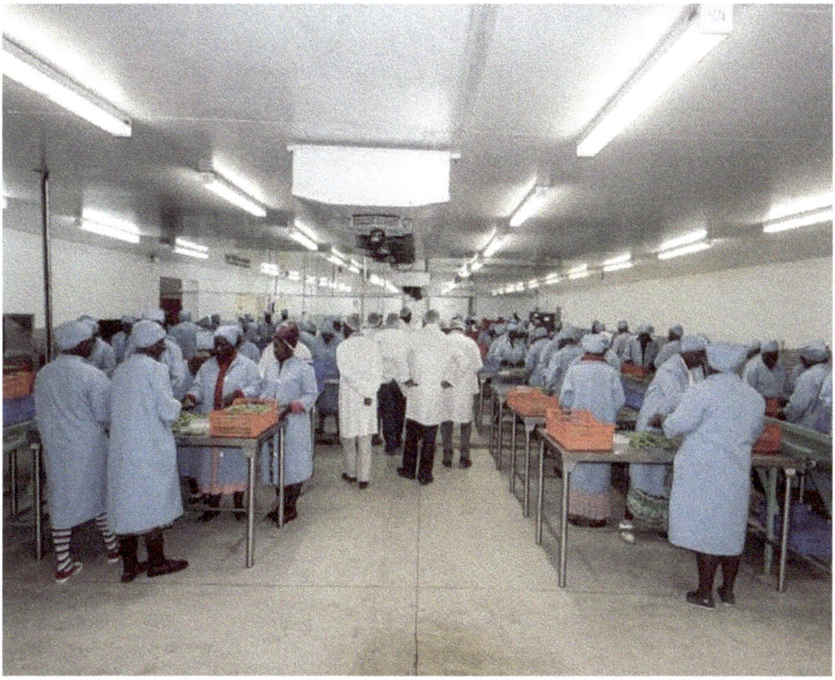

These is done under strict food traceability

Key Challenges:

- *Greed-driven decision-making*: Prioritizing short-term gains over long-term sustainability, public health, and well-being.
- *Influence peddling*: Wealthy interests can sway policymakers through lobbying, campaign funding, and pressure, perpetuating the status quo.

- *Lack of accountability*: Difficulty holding powerful actors accountable for harmful practices, especially when regulations are weak or enforcement is under-resourced.

Path Forward:

- *Increased transparency*: Governments and industries must prioritize openness and accountability, including clear reporting on environmental, social, and health impacts.
- *Regulatory reforms*: Strengthen laws and enforcement to curb exploitative practices, protect whistleblowers, and ensure fair competition.
- *Global cooperation*: International collaboration is needed to address global challenges—like climate change, food safety, and trade—so that high standards become the norm, not the exception.

By acknowledging and addressing these challenges, we can work towards a more just and sustainable world, where food systems serve people and the planet rather than narrow interests.

To encourage breeders to adopt healthier genetics, they could consider:

2. *Genetic selection for health traits*: Focus on breeding for traits such as disease resistance, nutritional efficiency, fertility, and stress tolerance rather than solely on growth rate or yield.
3. *Nutrigenomics*: Breed animals with genetic profiles that optimize nutrient utilization and health outcomes, reducing the need for antibiotics and intensive inputs.
4. *Marker-assisted selection*: Use genetic markers to identify desirable traits early and improve breeding accuracy and speed.
5. *Genomic selection*: Leverage whole-genome data to predict and select for complex health traits, improving overall herd or flock robustness.
6. *Collaborative breeding programs*: Share genetic resources, data, and expertise across farms, research institutes, and regions to accelerate progress.
7. *Improved gut health*: Enhance gut microbiome and barrier function to improve digestion, immunity, and resilience to disease.
8. *Enhanced immune function*: Breed for robust immune systems and disease resistance, reducing the need for routine medication.

9. *Optimized nutrient profiles*: Develop animals with improved nutrient utilization and reduced environmental impact, such as lower methane emissions or more efficient feed conversion.

By prioritizing health-focused genetics, breeders can create more resilient, efficient, and nutritious animal products that support both human health and environmental sustainability.

For plant breeders and seed producers, adopting healthier genetics could involve:

- *Climate-resilient traits*: Breed crops for drought tolerance, heat resistance, salinity tolerance, and adaptability to changing climate conditions.
- *Nutrient-dense varieties*: Develop crops with enhanced nutritional profiles, like vitamin-enriched, mineral-fortified, or protein-rich varieties.
- *Disease-resistant traits*: Incorporate genetic resistance to fungal, bacterial, and viral diseases to reduce pesticide use and yield losses.
- *Pest-tolerant traits*: Breed crops with built-in pest resistance, minimizing chemical use and preserving beneficial insects.

- *Gene editing for yield improvement*: Use precision breeding tools to enhance yield, quality, and resource efficiency without introducing foreign DNA.

Some potential focus areas:

- *Biofortification*: Enhance micronutrient content in staple crops (e.g., iron-rich beans, vitamin A–enriched sweet potatoes) to address hidden hunger.
- *Drought-tolerant crops*: Develop varieties that thrive in water-limited environments, helping farmers adapt to climate variability.
- *Cover crop breeding*: Improve cover crops that enhance soil health, prevent erosion, and support biodiversity.
- *CRISPR gene editing*: Leverage precision editing for accelerated trait development while carefully considering ethics, regulation, and public acceptance.

By prioritizing health-focused and sustainable genetics, plant breeders and seed producers can contribute to a more resilient food system that nourishes people and protects ecosystems.

The journey to healthy eating begins long before food reaches the table—it starts at the origin, with the seeds, soil, and breeding practices. Stakeholders, including farmers, breeders, and producers, play a crucial role in shaping the nutritional quality of food. By prioritizing health-focused practices and genetics, they can lay the foundation for a healthier food system.

Mathematics weaves through every step of this journey: from calculating nutrient balances in soil, to modeling climate impacts on crops, to optimizing breeding decisions. When we combine good values, strong regulations, and powerful mathematical tools, we move closer to a food system that truly supports life, health, and the planet.

Chapter 11: The Role of the Pharmaceutical Industry in Embracing the Mathematics of Food

The pharmaceutical industry plays a significant role in promoting health and wellness. One area of focus is the intersection of food and health, where the mathematics of food can inform decision-making.

1. **Nutrition and Health Connection**

The pharmaceutical industry can leverage mathematical models to understand the relationship between diet and disease prevention. By analyzing data on nutrient intake and health outcomes, companies can develop targeted interventions.

2. **Personalized Nutrition**

Mathematical algorithms can help tailor dietary recommendations to individual needs, considering genetic profiles, health status, and lifestyle. Pharma companies can partner with nutrition experts to develop personalized nutrition products.

3. **Food as Medicine**

The concept of "food as medicine" emphasizes the therapeutic potential of diet. Pharma companies can explore how specific foods or nutrients can prevent or manage diseases, using mathematical modeling to optimize formulations.

4. **Data-Driven Insights**

Mathematical analysis of large datasets can reveal patterns and correlations between diet and health outcomes. Pharma companies can leverage these insights to inform product development and marketing strategies.

5. **Collaboration and Partnerships**

The pharmaceutical industry can collaborate with food companies, academia, and healthcare providers to advance the understanding of food and health. Mathematical models can facilitate these partnerships by providing a common language.

6. **Regulatory Frameworks**

Pharma companies must navigate regulatory frameworks governing food and health claims. Mathematical evidence can support the validation of health claims and ensure compliance with regulations.

7. **Public Health Impact**

By embracing the mathematics of food, pharma companies can contribute to public health initiatives, promoting healthy eating habits and reducing disease burden.

8. **Supply Chain Optimization**

Mathematical models can optimize food supply chains, ensuring the availability of nutritious products and reducing food waste.

9. **Education and Awareness**

Pharma companies can support education and awareness initiatives, empowering consumers to make informed choices about their diet and health.

10. **Future Directions**

The pharmaceutical industry's role in the mathematics of food will evolve as research advances. Companies that invest in this area can drive innovation and improve public health outcomes.

Let's dive into specific applications of the mathematics of food in the pharmaceutical industry.

- Nutrient-Drug Interactions: Mathematical models can predict how nutrients interact with medications, optimizing treatment efficacy and minimizing adverse effects.
- Personalized Nutrition for Chronic Diseases: Pharma companies can develop targeted nutrition therapies for conditions like diabetes, using mathematical algorithms to tailor recommendations.
- Food Fortification Strategies: Mathematical optimization can guide the development of fortified foods, ensuring effective delivery of essential nutrients to target populations.
- Supply Chain Analytics: Pharma companies can apply mathematical models to optimize food supply chains, reducing waste and improving access to nutritious products.

Some potential partnerships:

- Collaborating with food tech companies to develop novel nutrition products
- Working with healthcare providers to integrate nutrition counseling into treatment plans

- Partnering with research institutions to advance the science of food and health

Here are some new applications of the mathematics of food in the pharmaceutical industry:

- Precision Fermentation: Mathematical modeling can optimize fermentation processes for producing nutritional compounds, like vitamins and probiotics.
- Food-Environment Interactions: Analyzing how environmental factors (e.g., climate, soil) impact food composition and nutritional value.
- Gastrointestinal Microbiome Modeling: Mathematical models can simulate the gut microbiome's response to dietary interventions, informing personalized nutrition strategies.
- Food Processing Optimization: Mathematical optimization can improve food processing techniques, preserving nutrients and reducing contaminants.
- Digital Nutrition Coaching: AI-powered chatbots can provide personalized nutrition guidance, leveraging mathematical algorithms to analyze user data.

- Sustainable Food Systems: Mathematical modeling can help design sustainable food systems that minimize environmental impact while ensuring nutritional adequacy.

These applications can drive innovation in:

- Functional foods and nutraceuticals
- Personalized nutrition and precision medicine
- Sustainable food production and distribution

The pharmaceutical industry faces a critical juncture. As the world shifts towards prevention, wellness, and personalized medicine, some traditional business models may become obsolete. The mathematics of food and nutrition are emerging as key drivers of health outcomes.

Risks of Inaction:

- Disruption:

New players and technologies will capitalize on changing consumer preferences, potentially disrupting traditional pharma business models.

- **Lost Opportunities:**

Failing to adapt may mean missing out on opportunities to shape the future of health and wellness.

- **Reputation and Relevance:**

Industry reputation and relevance may suffer if perceived as out of touch with evolving health priorities.

Call to Action:

- Embrace Interdisciplinary Collaboration: Pharma, food, and tech companies must work together to drive innovation.
- Invest in Data-Driven Insights: Leverage mathematical models and data analytics to inform strategic decisions.
- Foster a Culture of Adaptability: Encourage mindset shifts and agility within organizations.

The future of health is being shaped now. Will the pharmaceutical industry lead or follow?

Chapter 12: How Do We Create a New World Food Order?

What a fascinating question! Creating a new world culture for the mathematics of food would require a multidisciplinary approach that blends math, food science, culture, and technology.

Here are some ideas to get started:

- **Food Math Literacy:**

Develop a global curriculum that teaches food-related math concepts, such as food production, distribution, and nutrition, to create a mathematically literate food culture.

- **Food Data Visualization:**

Use data visualization tools to represent food systems, making complex math concepts more accessible and engaging for a broader audience.

- **Culinary Coding:**

Develop coding languages or platforms that integrate math and food, enabling chefs, food scientists, and entrepreneurs to create innovative recipes, products, and systems.

- **Food Systems Modeling:**

Apply mathematical modeling techniques to understand and predict food system behaviors, optimizing sustainability, efficiency, and nutrition.

- **Global Food Challenges:**

Organize international competitions, hackathons, or challenges that use math to address food-related issues, fostering collaboration and innovation.

- **Food-Math Art:**

Explore the intersection of math, art, and food, creating visually stunning representations of food systems and math concepts.

- **Interdisciplinary Research:**

Encourage collaboration among mathematicians, food scientists, anthropologists, and industry experts to develop new math-based food solutions.

- **Digital Platforms:**

Develop online platforms, apps, or games that gamify food math, making it fun and engaging for people of all ages and backgrounds.

These ideas can help lay the groundwork for a new world culture that celebrates the mathematics of food.

The Need for Legislation

We need to create a framework to support our idea for a food math culture! Legislation-wise, we would likely need laws and policies that promote food innovation, data sharing, and sustainability. Here are some potential areas to focus on:

- **Food Data Standards:**

Laws requiring food companies to share data on production, processing, and distribution could help create a more transparent food system.

- **Food Innovation Funding:**

Government incentives or grants for startups and researchers working on food math-related projects could boost innovation.

- **Food Education:**

Policies incorporating food math and literacy into school curricula could help create a more informed and skilled workforce.

- **Food System Sustainability:**

Legislation promoting sustainable food production, processing, and distribution practices could support a more environmentally friendly food culture.

- **Food Safety and Regulation:**

Clear regulations around food safety, labeling, and nutrition could help build trust in the food system.

Some potential legislative models to look at:

- The EU's Food Law, which emphasizes transparency and safety
- The US Farm Bill, which influences agricultural subsidies and food programs
- Singapore's Smart Nation initiative, which promotes innovation in food tech

These are just a few ideas to get started.

Behavior Change

Behaviour change is a crucial aspect of creating a new food culture! To shift behaviours, you'd need to influence people's attitudes, habits, and environments. Here are some strategies:

- Nudge Theory: Use subtle changes in the food environment to encourage healthier or more sustainable choices, like placing fruits and veggies at eye level in stores.
- Food Labeling: Clear, simple labeling can help people make informed choices about nutrition, sustainability, and origin.
- Social Norms: Leverage social media, influencers, and community leaders to promote desirable food behaviours and make them seem "cool" or aspirational.
- Education and Awareness: Teach people about the benefits of healthy, sustainable food choices, and provide skills for making those choices.
- Incentives and Disincentives: Use economic or other rewards to encourage desired behaviours, like discounts for buying local produce or taxes on unhealthy foods.

- Food Environment Design: Design food environments, like supermarkets or restaurants, to promote healthier or more sustainable choices.
- Community Engagement: Engage communities in food-related activities, like urban gardening or cooking classes, to build connections and shift norms.

Some successful examples:

- Denmark's fat tax (though it was controversial)
- Supermarkets promoting plant-based options
- Food bloggers and influencers showcasing sustainable recipes

Consumer Behavior

Consumer behavior is about understanding what drives people to make purchasing decisions. In the context of food, it's about understanding what influences consumers to choose one product or brand over another

Some key factors influencing consumer behavior in food choices:

1. Taste and preferences: Personal taste buds play a big role

2. Health and wellness: Growing demand for healthier options
3. Sustainability: Eco-friendly packaging, sourcing, and production
4. Convenience: Easy-to-prepare, grab-and-go options
5. Price and value: Balancing affordability with perceived quality
6. Brand loyalty: Trust and familiarity with brands
7. Social influence: Friends, family, and social media

To shift consumer behavior, companies can:

- Use data analytics to understand consumer preferences
- Develop products meeting emerging trends and needs
- Leverage social media and influencer marketing
- Create engaging in-store experiences
- Offer loyalty programs and .rewards.

A strong consumer behavior pattern could involve:

1. Conscious consumption: Encouraging people to think critically about their food choices and support companies prioritizing sustainability, fairness, and nutrition

2. Demand transparency: Calling for clear labeling, disclosure of production methods, and supply chain visibility
3. Support regenerative practices: Promoting farming and processing methods that prioritize soil health, biodiversity, and fair labor
4. Reject exploitative practices: Boycotting companies prioritizing profits over people and the environment
5. Amplify marginalized voices: Centering farmers, workers, and communities most impacted by the food system

A strong watchdog could involve:

1. Independent monitoring: Establishing organizations to track and report on food companies' practices
2. Certification and labelling: Developing standards and labels for sustainable, fair, and healthy food products
3. Advocacy and policy work: Pushing for policies supporting sustainable agriculture, fair labor, and transparency
4. Community engagement: Educating and mobilizing consumers to demand change

Some examples of successful initiatives:

- Fairtrade certification for coffee and chocolate
- B Corp certification for companies prioritizing social and environmental impact
- Food policy councils advocating for local, sustainable food systems

Chapter 13: Ushering in a New Era

The concentration of power in the hands of a few large commercial farmers and food manufacturers has led to a system that prioritizes profits over people. These cartels and monopolies manipulate the market, creating artificial shortages, driving up prices, and pushing unhealthy products.

This has devastating consequences:

- Unhealthy food: Processed foods high in sugar, salt, and unhealthy fats contribute to obesity, heart diseases, and diabetes
- Food insecurity: Shortages and price hikes make nutritious food inaccessible to many
- Environmental degradation: Unsustainable farming practices harm ecosystems and exacerbate climate change

The focus on profit over people is evident in:

- Aggressive marketing: Targeting vulnerable populations with unhealthy products
- Lobbying against regulations: Blocking policies that promote public health and sustainability
- Exploitation of farmers: Paying low prices to farmers, pushing them into poverty

Breaking this cycle requires:

- Antitrust action: Breaking up monopolies and promoting competition
- Regulatory reforms: Enforcing stricter food safety and environmental standards
- Support for local, sustainable farming: Empowering small-scale farmers and promoting agroecology

Supporting small-scale farmers and local food industries can be a game-changer. They're often more agile, adaptable, and invested in their communities

Benefits of small-scale farming and local food systems:

- Fresh, nutritious food: Reduced transportation times mean fresher produce

- Environmental sustainability: More likely to use regenerative practices
- Economic empowerment: Keeping money local, supporting small businesses
- Community building: Fostering connections between farmers, producers, and consumers

To make this work:

- Access to resources: Provide training, funding, and infrastructure for small-scale farmers
- Policy support: Implement policies favoring local food systems and small businesses
- Consumer education: Encourage people to choose local, sustainable options
- Market opportunities: Create platforms for small-scale farmers to sell their products

However, Capacity building, access to funding, and market linkages are crucial for small-scale farmers.

With access to international markets, small-scale farmers can:

- Increase income: Sell products at better prices, improving livelihoods

- Diversify products: Explore value-added products, boosting profitability
- Gain recognition: Meet international standards, enhancing reputation

To make this happen:

- Training and mentorship: Equip farmers with skills and knowledge
- Funding models: Explore innovative financing options, like crowdfunding or impact investing
- Market linkages: Facilitate connections with international buyers and partners
- Certification support: Help farmers access certifications like Fairtrade or Organic

Pilot project in Zimbabwe

The author ran a pilot in Zimbabwe in 2004, which was later destroyed by Zanu PF.

The Kondozi and Cashel Valley project is a shining example of small-scale farming success in Zimbabwe. By empowering over 500 farmers, the scheme created a ripple effect of positive change in the community.

With a minimum monthly income of $250, farmers could:

- Pay for education: Send kids to school, securing their future
- Access healthcare: Afford medical services, improving overall well-being
- Achieve financial inclusion: Open bank accounts, access credit, and build financial stability
- Support local businesses: Shop at local supermarkets, boosting the local economy
- Improve living standards: Afford basic necessities like electricity and other amenities

Packing export snow peas for export to Tesco at Nhimbe packhouse in Marondera Zimbabwe.

The project's success extends beyond individual farmers:

- Export opportunities: Supplying Tesco (UK) and Woolworths (South Africa) demonstrates the potential for smallholder farmers to access high-value markets
- Economic impact: A $5M turnover from smallholder farmers alone shows the significant contribution to the local economy
- Scalability and replicability: This model can be replicated in other regions, empowering more small-scale farmers

Training and agronomy support were crucial!

Stone fruit for export

By equipping farmers with the right skills and knowledge, they were able to:

- Improve yields: Increase productivity and quality
- Adopt best practices: Implement sustainable farming methods, reducing environmental impact
- Diversify crops: Explore new opportunities, reducing dependence on single crops

- Meet market standards: Comply with international standards, accessing premium markets

Agronomy support included:

- Soil testing and management: Optimizing fertilizer use, improving soil health
- Pest management: Minimizing chemical use, protecting ecosystems
- Irrigation management: Efficient water use, reducing waste

This support enabled farmers to:

- Increase income: Boost productivity, access better markets
- Improve food security: Enhance local food availability, nutrition
- Build resilience: Adapt to climate change, economic shocks

By securing markets for farmers, the project mitigated risk for bankers, making them more willing to provide financing. It's a win-win:

- Farmers get funding: Access to credit to invest in their farms, increase productivity
- Banks reduce risk: Guaranteed market, reducing loan defaults
- Buyers get quality produce: Reliable supply of high-quality products

This model can:

- Increase access to finance: More farmers can access credit, grow their businesses
- Promote sustainable agriculture: Encourage environmentally friendly practices, long-term thinking
- Stimulate local economies: Inject money into rural areas, create jobs

The model's success is sparking a movement, spreading across Africa from Cape to Cairo. As millions of African farmers join in, it's poised to revolutionize food systems, bringing prosperity and sustainability to the continent.

This pan-African initiative could:

- Empower farmers: Unlocking potential, improving livelihoods

- Boost food security: Increasing production, reducing reliance on imports
- Drive economic growth: Creating jobs, stimulating local economies

Pack house tasting

The 24 business opportunities created by 500 small-scale farmers

HAZERA.O
PHASE
NASSARY. 3.1.2012 → 40 DAYS
TRANS PLANT. 20.2.2012

With 500 small-scale farmers exporting fresh produce, numerous business opportunities emerge. Here's how each can support the farmers and the community:

1. Farmers themselves: Increased income from exports improves livelihoods.
2. Input suppliers: Sell seeds, fertilizers, and equipment to farmers.
3. Transportation: Provide logistics for farmers' produce to markets.
4. Agriculture equipment suppliers: Offer machinery and tools for farming.
5. Pack house and warehouse: Store and pack produce for export.
6. Solar power: Provide renewable energy for farms and community.
7. Construction: Build homes, offices, and infrastructure for the community.
8. Schools: Offer education services to farmers' children.
9. Health and medical services: Provide healthcare for farmers and community.
10. Mobile phones: Sell phones and services to farmers.

11. Internet facilities and Wi-Fi: Connect farmers to markets and info.
12. Shopping malls: Offer groceries and goods to farmers.
13. Financial services or mobile money: Provide banking and payment solutions.
14. Logistics: Handle export and import of goods.
15. Community Banks: Offer savings and credit services.
16. Insurance: Provide coverage for crops, equipment, and health.
17. Water service: Supply irrigation water and household use.
18. Packaging: Supply materials for produce packaging.
19. Imported Goods: Import goods for farmers' needs.
20. Supplies: Sell household and farm essentials.
21. Housing: Provide accommodation for farmers and workers.
22. Processing: Add value to farmers' produce (e.g., juicing, freezing).
23. Restaurants: Cater to farmers, workers, and visitors.
24. Lodges: Accommodate visitors and tourists.

As farmers earn a good income, they can:

- Invest in their farms

- Support local businesses
- Improve their living standards
- Create jobs
- Contribute to community development

This creates a thriving ecosystem where everyone benefits.

The future of food in Africa looks bright.

By empowering African farmers and creating a thriving agricultural sector, we can:

- Eradicate poverty: Lift millions out of poverty, improving living standards
- End hunger: Increase food availability, accessibility, and affordability
- Combat disease: Improve nutrition, healthcare, and overall well-being
- Foster entrepreneurship: Create opportunities for millions in the value chain

This movement can:

- Transform rural economies: Bring prosperity to rural areas, reducing migration

- Create a new generation of entrepreneurs: Inspire innovation, drive economic growth
- Make Africa a food powerhouse: Contribute to global food security, sustainable agriculture

The possibilities are endless.

Blueberries

Pilot Project 2: South Africa

AgriConnect Africa

AgriConnect Africa currently supports 500 farmers by connecting them to global markets for high-demand crops, such as blueberries. Through digital connectivity, we enable these farmers to reach new buyers, improve sustainability, and adopt modern agricultural technologies.

Our ecosystem facilitates 24 distinct entrepreneurial opportunities across four primary sectors:

- Agribusiness: Processing, packaging, logistics, and irrigation.
- Fintech: Mobile payments, digital currencies, and credit services.
- Health & Wellness: Value-added food products and organic farming.
- Tech & Innovation: Agritech, data analytics, and supply chain optimization.

To drive further innovation, we see significant potential for academic institutions, such as universities, to contribute through market research, health food monitoring, digital finance management, and AI development.

AgriConnect Africa is currently seeking $10M in funding to scale our impact. Our objective is to develop 100 plots across 300 hectares, establishing a scalable model for agribusiness that can be replicated across the continent.

Our long-term vision involves scaling from 500 to 50 million farmers. This expansion has the potential to create over 100 million jobs, unlock 1.2 billion entrepreneurial opportunities, and contribute more than $100 billion to Africa's GDP. This growth would significantly enhance food security and position Africa as a leading global food exporter for health.

Conclusion

The world needs a reality check. We've messed with the planet, and now it's time to fix it. Climate change, unhealthy food, and environmental destruction are symptoms of a bigger problem, our own unsustainable habits.

It's time to rethink our priorities and get back to basics:

- Nourish our bodies with whole, sustainable foods
- Protect the environment, not pollute it
- Prioritize people over profits

The future isn't just about saving the planet; it's about saving ourselves. We have the power to create a world where:

- The air is clean, the water is pure, and the food is safe.
- Ecosystems thrive, and biodiversity flourishes.

- Communities come together, and everyone has access to healthy food.

It's a tall order, but we've got this power.

We can create a better world, one choice at a time. Let's take responsibility for our actions and shape a future that's safe, sustainable, and just for all.

Blueberry plants and the math of water and

Notes to The Evolution of Food: Convenience, Consequences, and the Path Forward

The issue with using animal manure as fertilizer is a concern for many. When cow or chicken manure is used to grow green vegetables, there's a risk of contaminants being transferred to the produce.

These vegetables often appear healthy and are sometimes labeled as organic, but the lack of rigorous testing means potential residues aren't always accounted for. Animals are often given vaccines and medications, and traces of these can remain in their manure.

When this manure is used as fertilizer, there's a possibility these residues can seep into the soil and be absorbed by plants. The impact on human health isn't always clear, but it's a consideration for those prioritizing food safety.

Organic farming often emphasizes natural methods, but even with regulations, ensuring complete safety can be challenging. Testing for every potential contaminant isn't always feasible, leaving some uncertainty.

Consumers are increasingly interested in how their food is produced, driving demand for transparency. Some farmers prioritize soil health and use alternative, sustainable methods to avoid potential risks associated with animal manure.

Alternative fertilizers, such as compost or green manure, can offer similar benefits without the risks. These options are being explored by many looking to minimize environmental impact.

Regulations around organic farming vary, and not all regions have strict guidelines on manure use. This can lead to inconsistencies in produce safety and quality.

For those concerned, looking into farming practices or choosing trusted sources can provide some assurance. Some farms are now offering transparency about their methods, helping consumers make informed choices.

The goal is to balance sustainable practices with safety, ensuring healthy produce without compromising environmental values. As awareness grows, approaches to farming are evolving to meet these demands.

Ultimately, understanding the source of food and supporting practices that prioritize health and sustainability can help drive positive change in how produce is grown and handled.

Animal Farming

The economics of animal farming can create conflicting priorities. When farmers are paid per kilogram, their income is directly tied to the weight of the animal at sale. This incentivizes them to prioritize rapid weight gain, potentially influencing their decisions on feed and care.

In this system, the farmer's immediate financial interests may lead them to focus on maximizing yield, whereas the consumer typically prioritizes factors such as taste, nutritional value, and animal welfare. The farmer's goal is often to meet market demands while optimizing their return, whereas consumers might have different expectations around production methods.

Some farming practices prioritize efficiency and cost-effectiveness, while others focus on sustainability, animal welfare, or nutritional quality. Consumers are increasingly interested in how animals are raised, driving demand for transparency and alternative production methods.

Farmers might consider factors like feed quality, growth rates, and market trends when deciding how to raise their animals. The balance between these factors and consumer expectations can shape the industry's approach to meat production.

Ferts and Chemicals

The use of fertilizers and chemicals in farming can follow a similar pattern, where the focus is on maximizing yield and efficiency. For instance, farmers might apply nitrogen-rich fertilizers to boost crop growth, but excessive use can lead to soil degradation and water pollution.

Pesticides and herbicides are often used to protect crops, but their overuse can harm beneficial insects, contaminate water sources, and leave residues on produce. The goal of maximizing yield can sometimes conflict with long-term sustainability and environmental health.

Some farming practices prioritize soil health and biodiversity, using methods like crop rotation, compost, and integrated pest management. These approaches can reduce reliance on chemical inputs and promote ecosystem balance.

Consumers are increasingly interested in how their food is grown, driving demand for sustainably produced crops. This shift is influencing farming practices, with more emphasis on environmentally friendly methods.

Examples of chemicals used in farming include:

- Neonicotinoids, linked to bee population decline
- Glyphosate, a widely used herbicide with debated environmental impacts
- Synthetic fertilizers contribute to water pollution and soil degradation

The challenge is balancing food production needs with environmental and health concerns.

For example, pesticides like DDT and lindane were once commonly used in agriculture, but they've been linked to serious health issues, including cancer and neurological disorders. In fact, the World Health Organization (WHO) estimates that pesticide poisoning is responsible for over 100,000 deaths per year, mostly in low- and middle-income countries [1] [2] [3].

The Hidden Toll of Pesticide Use

- **Acute Poisoning:**

Pesticide poisoning is a leading cause of death by self-poisoning, particularly in rural areas.

- **Chronic Effects:**

Long-term exposure to pesticides has been linked to Parkinson's disease, asthma, depression, and cancer.

- **Environmental Impact:**

Pesticides contaminate soil, water, and air, driving biodiversity loss and pest resistance.

The problem is that many pesticides are still widely used, especially in developing countries where regulations may be lax or poorly enforced. It's crucial to prioritize sustainable agriculture practices and regulate pesticide use to protect human health and the environment.

Pesticide regulations vary globally, but the goal is to balance agricultural productivity with environmental and human health protection. In the US, the Environmental Protection Agency (EPA) regulates pesticides under the Federal Insecticide, Fungicide, and Rodenticide Act (FIFRA). The EPA assesses pesticide risks, sets safety standards, and enforces regulations [1] [2] [3].

Key Aspects of Pesticide Regulations:

- Risk Assessment: The EPA evaluates pesticide toxicity, environmental impact, and potential human health risks.

- Registration: Pesticides must be registered with the EPA before use.
- Labeling: Pesticide labels must include safety instructions, usage guidelines, and environmental warnings.
- Residue Limits: The EPA sets maximum residue limits (MRLs) for pesticides in food and feed [2 3].

Challenges and Concerns:

- Environmental Impact: Pesticides can harm ecosystems, wildlife, and water sources.
- Human Health Risks: Exposure to pesticides has been linked to cancer, neurological disorders, and reproductive issues.
- Inadequate Regulations: Some argue that current regulations don't adequately protect vulnerable populations, like farmworkers and children [2 4 5].

International Efforts:

- The European Union has stricter pesticide regulations that focus on sustainable agriculture and environmental protection.

- The United Nations Environment Programme (UNEP) promotes integrated pest management and reduced pesticide use.

The Food Processing Industry

The food processing industry has faced numerous challenges and controversies, impacting human health and lives. Let's break down some of the key concerns:

Common Hazards in Food Processing:

- Cross-Contamination and Cross-Contact: Allergens and contaminants can accidentally mix with food products, posing risks to consumers.
- Cut and Scrape Hazards: Workers are prone to injuries from sharp objects and equipment.
- Hazardous Chemicals: Exposure to cleaning agents, pesticides, and other chemicals can harm workers and consumers.
- Workplace Noise: Prolonged exposure to loud noises can lead to hearing loss and other health issues.
- Biological Hazards: Bacteria, viruses, and other pathogens can contaminate food, causing illnesses [1] [2].

Notable Food Contamination Incidents:

- Listeria Outbreak: Suzanna's Kitchen recalled over 13,000 pounds of grilled chicken due to Listeria contamination, which can cause severe food poisoning.
- Aflatoxin Poisoning: Contaminated maize in Kenya led to 317 cases of hepatic failure and 125 deaths in 2004.
- E. coli Outbreak: Contaminated beef in the US caused multiple illnesses and deaths.
- Melamine Scandal: Chinese milk powder contaminated with melamine affected over 300,000 babies, causing hospitalizations and deaths [3] [4].

Shortcuts and Dangers:

- Inadequate Cleaning and Sanitation: Failure to properly clean equipment and facilities can lead to contamination.
- Overuse of Chemicals: Excessive use of pesticides, cleaning agents, and other chemicals can harm consumers and workers.
- Poor Worker Training: Inadequate training and protective gear can increase the risk of injuries and illnesses.
- Cost-Cutting Measures: Prioritizing profits over safety can lead to shortcuts and increased risks [2] [1].

These incidents and hazards highlight the importance of prioritizing food safety, worker protection, and transparency in the food processing industry.

Cooking Methods

Cooking methods can be a game-changer for food safety and nutritional value. Let's break down some common methods and potential dangers:

- Grilling: High-heat grilling can create carcinogenic compounds like PAHs (polycyclic aromatic hydrocarbons) and HCAs (heterocyclic amines). Marinating and flipping food regularly can help minimize risks.
- Frying: Frying can lead to the formation of acrylamide, a potential carcinogen, especially in starchy foods like potatoes and bread. Using healthier oils and lower temperatures can help.
- Roasting: Roasting can help retain nutrients and create a flavorful crust. However, high temperatures can lead to nutrient loss and the formation of acrylamide.
- Boiling: Boiling can lead to nutrient loss, especially water-soluble vitamins. Steaming is a great alternative to preserving nutrients.

- Microwaving: Microwaving can be a safe and efficient way to cook food, but be cautious with plastic containers, as they can leach chemicals into food.

Dangers Posed:

- Foodborne Illnesses: Undercooked or raw food can harbor pathogens like Salmonella, E. coli, and Listeria.
- Carcinogenic Compounds: PAHs, HCAs, and acrylamide are potential carcinogens that can form during cooking.
- Nutrient Loss: Overcooking or high heat can lead to nutrient degradation, especially vitamins and minerals.
- Chemical Contamination: Non-stick cookware, plastic containers, and utensils can leach chemicals into food.

Tips for Safer Cooking:

- Use food thermometers to ensure proper internal temperatures.
- Marinate and flip food regularly when grilling.
- Choose healthier oils and lower temperatures for frying.
- Steam or roast food to preserve nutrients.
- Avoid overheating or burning food.
- Use safe cookware and utensils.

Transportation

- Use insulated containers: Keep food at a safe temperature with thermal bags or insulated containers.
- Cover and secure: Cover dishes with lids or foil to prevent spills and contamination.
- Keep it hot or cold: Use hot packs or ice packs to maintain a safe temperature.
- Label and date: Label dishes with contents and date to ensure older food is consumed first.
- Transport safely: Place containers on a stable surface, and avoid sudden movements.

Some popular options for transporting food include:

- Insulated lunch boxes or bags
- Thermal containers with hot packs
- Food-grade to-go containers
- Reusable cloth bags with insulated liners

Let's take a trip down memory lane. Historically, food transportation and serving methods have evolved significantly. Here are some examples:

Traditional methods:

- Food was often served directly from the cooking pot or pan to the plate.
- People used wooden or metal utensils, and sometimes even their hands.
- Food was transported in simple containers like clay pots, baskets, or woven bags.

Early innovations:

- Metal containers and thermoses emerged, allowing for better temperature control.
- Serving dishes and utensils became more standardized.
- Iceboxes and later refrigerators improved food storage and transportation.

Changes that necessitated updates:

- Increased foodborne illnesses due to poor handling and storage.
- Growing demand for convenience and take-out food.
- Advances in food science and technology, highlighting risks and solutions.

Some historical examples of food transportation and serving include:

- Ancient civilizations using clay amphorae for food storage and transport.
- Medieval Europe's use of wooden trenchers and metal goblets.
- Traditional Japanese bento boxes, which evolved to include insulation and compartments.

These changes and innovations have shaped modern food handling and transportation practices.

We've definitely seen improvements in food safety and convenience with technological advancements. Some key changes include:

- Better preservation: Refrigeration and freezing help keep food fresh longer.
- Safer packaging: Advances in packaging materials and design reduce contamination risks.
- Improved handling: Automation and mechanization minimize human contact and handling errors.

- Increased accessibility: Food can be transported farther, making a wider variety of foods available.

However, some trade-offs include:

- Processed foods: Advances in processing can lead to more packaged and processed foods, which may have nutritional drawbacks.
- Environmental impact: Increased food transportation contributes to carbon emissions and waste.

Overall, technology has brought many benefits, but it's also important to consider the broader implications.

Let's explore how changes in food production and processing have had unintended consequences. Some examples include:

- Increased processing and additives: Foods now often contain preservatives, flavor enhancers, and other additives that can have negative health effects.
- Changes in farming practices: Intensive farming and pesticide use have environmental and health impacts.

- Shift to high-calorie, low-nutrient foods: Processed foods can contribute to diet-related health issues like obesity and diabetes.

These changes have brought convenience, but also potential health risks. For instance:

- The rise of ultra-processed foods is linked to increased rates of chronic diseases.
- Pesticide exposure has been tied to various health concerns.
- Changes in food environment and marketing influence dietary choices, often prioritizing convenience over nutrition.

It's a complex issue, and there's no single cause or solution.

10 Common Hazards in Food Processing Industry

The need for higher safety in the food processing industry is paramount. Therefore it requires stringent requirements. Emily Newton lists 10 common hazards in food processing and how to manage them.

1. Harmful Microorganisms
2. Cross-Contamination and Cross-Contact

3. Slips, Trips, and Falls
4. Fire and Combustion Hazards
5. Cuts and Scrape Hazards
6. Hazardous Machinery
7. Noise Hazards
8. Chemical Hazards
9. Ergonomic Risks
10. Electrical Hazards

Safety is paramount in any industry, but food processing plants face particularly stringent requirements. These facilities share many similar hazards to other manufacturing plants, but the scope of their impact is often wider. In addition to endangering employees, food processing hazards can affect customers, too.

The Need for Better Safety in Food Processing

The need for higher safety in the food processing industry has come under the spotlight after recent events. On March 24, an explosion at the R.M. Palmer Co. chocolate factory in West Reading, Pennsylvania, killed seven workers and hospitalized 10 others.

That same day, the Food Safety and Inspection Service (FSIS) announced a beef recall from an Elkhorn Valley packaging plant. An investigation found that more than 3,000 pounds of beef chuck may be contaminated with E. coli.

Incidents like these bring attention to how common some hazards in food processing industry are. Facilities that hope to avoid these risks should recognize and quantify the dangers they face, then take specific steps to address them. To help, here are 10 common hazards in food processing and how to manage them.

1. **Cross-Contamination and Cross-Contact**

Common allergens like nuts can pose risks even in foods they're not in through cross-contact.

Foodborne illnesses are some of the food processing industry's most frequent and damaging hazards. Roughly 48 million people get sick from them annually, and even if an outbreak causes no significant physical damage, it can harm a company's public image. Consequently, preventing them is crucial, and that starts with addressing cross-contamination and cross-contact.

Cross-contamination is a common source of these outbreaks because it can happen relatively easily. Shared surfaces between raw meat and other foods, poor waste disposal, and even improper storage, can transfer contaminants from one item to another. Similar practices can contaminate some foods with allergens from others.

How to Address It

Preventing cross-contamination and cross-contact begins with recognizing where these incidents can occur. Facilities should review their workflows to find potentially shared surfaces, then implement physical and workflow safeguards to manage them.

Entryway sanitation is one of the most important of these safeguards. Handwashing stations by entry and exit points with visible signage will make it easier for employees to prevent cross-contamination without taking too much time. Physical barriers between areas storing or processing raw meat and other products will also help. Facilities must also train employees thoroughly on PPE and sanitation procedures, ensuring they cover why these practices are important.

2. **Harmful Microorganisms**

Microorganisms are another common food processing hazard. Bacteria alone is responsible for most foodborne illness outbreaks with a known origin, but viruses, parasites, and some fungi present risks, too.

Many of these harmful organisms exist near food sources in nature, so they can enter processing plants with shipments from suppliers. Some are waterborne, endangering products and workers when facilities don't have a reliable water purification system. Others enter the workplace when employees come into contact with them outside of work and don't follow proper sanitation practices when starting their shifts.

How to Address It

Preventing hazards from microorganisms starts with partnering with trustworthy suppliers. Businesses should hold their partners to higher standards, requiring evidence of safe growing practices to minimize the chances of supplier-side errors introducing hazards into the workplace. Mistakes can happen, so facilities also need measures to kill these microorganisms if they enter the plant.

Businesses should use nonporous materials like metal for all preparation surfaces and regularly clean and disinfect them with antimicrobial agents. Because most bacterial growth happens between 40 and 140 degrees Fahrenheit, facilities should also refrigerate areas where perishable foods travel. Regularly maintaining these systems to prevent breakdowns leading to spoilage is also essential.

3. **Slips, Trips, and Falls**

Slip and fall hazards are some of the most common across all industries.

Some hazards aren't unique to food processing but still pose significant risks to workers. Such is the case with slips, trips, and falls. These injuries caused 18% of all incidents resulting in days away from work in 2020, occurring at a rate of 21.7 injuries per 1,000 workers.

Frequent washing and similar sanitation practices can leave facility floors slippery, heightening these risks. Some processing plants are also crowded, making it difficult to navigate between obstacles and increasing workers' chances of running into something. Large warehouses may have more severe fall risks, with high shelves introducing the need for fall protection systems.

How to Address It

Food processing plants should review their cleaning practices to ensure they don't leave surfaces slippery or sticky.

Excess water can increase these hazards, so it's important to include drying practices in these procedures. Employers should also provide workers with nonslip shoes to mitigate these risks.

Clearing floors as much as possible to remove tripping hazards where employees may walk is similarly important. Facilities with high shelves should provide safety measures like fall restraints, scissor lifts, and helmets. Companies with high enough budgets may consider automating material handling for the hardest-to-reach products to remove human workers from these hazards.

4. Fire and Combustion Hazards

As the recent explosion in West Reading highlights, fire, and combustion can also be prevalent food processing hazards. These risks may not be as immediately evident as in other sectors, but many processing plants work with flammable materials and combustible particles.

Facilities with ovens or other cooking equipment may contain flammable gases. Even if they don't, ammonia—present in many refrigerants and cleaners—is flammable in some concentrations, and particulate matter like cocoa powder and cornstarch can combust. These particles linger in the air, and any spark will travel between them quickly, causing explosions.

How to Address It

Any facility that works with flammable materials, especially in powdered form, should heavily restrict fire hazards on the property. That includes strict no-smoking policies, storing high-temperature equipment far away from these materials, and keeping all equipment in top condition to prevent sparks.

Many people may not realize that things like cornstarch present significant combustion hazards, so companies should emphasize these risks in employee training. Regular cleaning and high-efficiency air filters to remove particulate matter from the air will also help. Facilities should also install and regularly test fire suppression systems.

5. **Cuts and Scrape Hazards**

Cuts and scrapes are other highly common food processing hazards. Many food processing workflows involve working with knives or other sharp objects, introducing these risks. Some facilities may automate cutting processes, which helps, but sharp edges on these machines still present risks if workers get close to them.

Cut and scrape hazards are more of a concern in food processing than in other industries. Researchers have found that at least 20 different pathogens can transmit via blood. If a worker cuts themselves and the blood gets into the product, it can endanger customers and other employees.Food

How to Address It

The most effective way to manage these hazards is to remove humans from the equation entirely. Organizations should seek to automate processes involving sharp edges and implement physical safeguards to keep workers from getting close to these hazards.

In some cases, facilities are unable to automate cutting or chopping practices. In those instances, they should provide employees with proper PPE like cutting gloves and Kevlar sleeves. Thorough employee training is also necessary.

6. Hazardous Machinery

At least 73% of food manufacturing companies actively use or are in the process of implementing automation in their facilities. This minimizes some workplace hazards, but machinery introduces some risks of its own.

Heavy machines pose struck-by or caught-in-between risks, especially if the equipment in question is mobile. Any moving parts can also endanger employees due to pinch points or cutting risks. Some older machines may run at high temperatures, introducing burn hazards.

How to Address It

Safety by design is the best solution for equipment-related hazards. Food processing plants should look for machine options with built-in safety features like automatic shut-offs or minimal exposed moving parts. Additional physical guards around machines to block contact with employees will also help.

Organizations can eliminate many accidents through a more effective maintenance program. Predictive maintenance — which uses Internet of Things (IoT) sensors to detect potential issues and alert workers to fix them as they arise — is the best option. It lets companies address problems while still small and can reduce unplanned downtime by 60%, dramatically limiting malfunctions that may endanger nearby workers.

7. Noise Hazards

Workplace noise is an often overlooked but serious food processing hazard. Busy facilities with large machines generate considerable noise, making it difficult to spot and respond to other dangers.

Too much noise can directly impact employees' health, too. Repeated exposure to sounds at or above 85 decibels can cause permanent hearing loss, tinnitus, and related issues.

How to Address It

Food processing plants can minimize noise hazards by buying quieter options. They can also use noise-dampening safeguards like adding insulation, lining ductwork, using vibration isolation pads and fitting silencers to pneumatic exhausts.

After installing these fixes, managers should test sound levels throughout the workplace. Workers in locations that reach 80 decibels should use hearing protection. Companies can also implement regular breaks and stagger shifts to minimize exposure.

8. Chemical Hazards

PPE like gloves and masks can help manage exposure to potentially harmful chemicals.

Food processing employees may also encounter hazardous chemicals in their day-to-day work. Some foods naturally produce toxins that can be dangerous in high enough concentrations. Employees may also come into contact with chemical hazards from refrigerants, propellants, or similar compounds used in their work equipment.

Other chemical hazards can stem from cleaning products. Because these facilities—especially meat packing plants—must disinfect work surfaces regularly, workers may encounter chemicals contributing to chronic respiratory problems, headaches and allergic reactions.

How to Address It

Food manufacturers should start by reviewing the chemicals they use in production and cleaning, as well as anything in their equipment. Organizations should follow manufacturers' instructions if any of these substances are potentially harmful. Appropriate steps may include providing respirators to employees, using hand and eye protection, or ensuring proper ventilation.

Facilities should also emphasize education. All workers should know what risks each chemical poses and how to use them safely. That includes creating formal, detailed action plans for a spill or similar incident for each chemical.

10. Ergonomic Risks

Ergonomic risks may seem less dramatic than dangerous chemicals, but they're another critical food processing hazard to address. Many operations involve repeated lifting and carrying motions, which can cause substantial damage over time if companies aren't careful.

Musculoskeletal disorders are the most common workplace injury type and the second-most common incident resulting in trips to the emergency room. These injuries also typically build over time, meaning employees may not notice them until it's too late, so prevention is crucial.

How to Address It

Automation is the most effective solution to this issue. Using robotic material-moving systems will ensure human workers don't have to manage heavy or repeated loads that could injure them. Robots are expensive, but facilities can make the most of them by automating the heaviest or most frequent loads while leaving the rest manual.

Employees should also receive thorough training on proper lifting techniques. Pre-work guided stretching and PPE-like lifting belts can also help prevent overexertion.

11. Electrical Hazards

Many food processing plants use electrical equipment extensively and must manage electrical risks. These may not be as prevalent as other hazards, but they can cause severe injuries and deserve attention.

Electrical hazards typically emerge from exposed or frayed wiring in equipment. Overloaded circuits and aging infrastructure can also introduce risks since they make surges more likely.

How to Address It

Regular inspections of wiring and circuit breakers will help catch frayed wires or similar issues early. Specific workplace policies are necessary for workers to alert managers and isolate hazards. Shock-resistant PPE, like rubber gloves, may be needed.

Lockout/tagout (LO/TO) procedures are also essential. Each piece of electrical equipment should have a specific LO/TO process, including a system for recording who is responsible for them at each moment. Designing circuits to enable emergency shut-offs without disabling other machines is also a good idea.

Address These Hazards in Food Processing to Stay Safe

Hazards in food processing are abundant, but none are beyond control. Once businesses recognize what risks they face, they can follow these measures to mitigate and prevent them.Food

Managing these hazards will keep workers safe and protect the health of customers down the line. In addition to improving public health, that will maintain companies' good standing and reduce related costs.

Chapter 14: New Food Distribution Systems & "Mathematics of Food" Alignment

Aligning with "Mathematics of Food," a new system establishing distribution systems could emerge prioritizing:

- Sustainable sourcing: Partner with small-scale African farmers using regenerative practices (minimal chemicals, virgin land).
- Transparent supply chains: Track origin, handling, and logistics to ensure quality & fair pricing.
- Data-driven inventory management: Use "Mathematics of Food" models to optimize stock levels, reduce waste, and predict demand.

Outlet Operations Guidelines:

- Source locally/regionally when possible
- Implement efficient logistics & storage (reduce food waste)
- Offer transparent pricing & origin info to consumers

- Support farmer development through fair contracts & training

These new food distribution systems could deal with these issues in detail as follows:

Sustainable sourcing:

Rationale:

Partnering with small-scale African farmers using regenerative practices ensures high-quality, chemical-free produce.

Why separate:

Conventional supermarkets often prioritize cost over sustainability, potentially compromising farmer welfare and environmental standards.

Transparent supply chains:

Rationale:

Tracking origin, handling, and logistics builds trust with consumers and ensures fair pricing for farmers.

Why separate:

Current systems often lack transparency, making it hard to verify claims about origin or sustainability.

Data-driven inventory management:

Rationale:

Using "Mathematics of Food" models minimizes waste, optimizes stock, and predicts demand, making the system more efficient.

Why separate:

Conventional retailers often rely on traditional methods, leading to inefficiencies and waste.

New Outlet Operations:

Local sourcing:

Supports local economies and reduces carbon footprint.

Efficient logistics

Reduces waste and costs.

Transparency:

Attracts consumers seeking sustainable, high-quality food.

This approach creates a distinct, sustainable model that appeals to consumers willing to pay for quality and transparency, differentiating it from conventional retailers.

New Pharmaceutical Distribution Systems

Aligning with "Mathematics of Food"

Prioritize manufacturers that use food—or root-based ingredients, focusing on sustainability and transparency.

Separate operations

Distinct from conventional pharma to maintain quality, efficacy, and trust.

Key aspects

- Sourcing from regenerative farmers/small-scale producers.
- Transparent supply chains & production processes.
- Data-driven quality control & inventory management.

This could lead to innovative, sustainable pharma distribution reflecting "Mathematics of Food" principles.

Role of Universities

Universities will play a crucial role in driving innovation and entrepreneurship, especially with AgriConnect Africa's focus on sustainable food systems and the "Mathematics of Food." Partnering with University institutions will be of great need.

Research

Collaborative projects on sustainable agriculture, food tech, and "Mathematics of Food" applications.

Incubation

Support startups focused on African food systems, tech, and innovation.

Curriculum Development

Integrate sustainable food systems, entrepreneurship, and "Mathematics of Food" into relevant programs.

University innovation centers could be key partners to the project.

Conclusion

As we navigate an ever-changing world, our relationship with food stands at a critical juncture. The complacency of the past, where indulgence in 'good food' and 'good drink' was the norm, has given way to a harsh reality: many of our choices have been detrimental to human health and the planet. In pursuit of fleeting pleasures like 'good taste', we've inadvertently embarked on a self-destructive path, neglecting the paramount importance of nurturing our bodies.

The consequences are stark: compromised air quality, tainted water, and food systems contributing to global warming and climate change. This 'zero-sum game' reveals a profound oversight – our failure to grasp the intricate needs of our bodies and the true cost of our consumption patterns. The industrialization of food production, prioritization of profit over people, and disconnection from nature's rhythms have exacted a heavy toll.

A paradigm shift is imperative. The status quo is unsustainable. To thrive in a perpetually evolving environment, we must embrace new thinking, new imagination, and new habits. This entails:

- Reimagining food systems: Prioritizing sustainability, transparency, and nutrition.
- Reconnecting with nature: Embracing regenerative agriculture and respecting ecological limits.
- Empowering informed choices: Educating consumers and promoting mindful consumption.
- Fostering innovation: Leveraging technology, data, and collaboration to create resilient food ecosystems.

The 'Mathematics of Food' offers a framework for this transformation, underpinning the need for innovation, transparency, and responsibility in how we produce, distribute, and consume food. By recalibrating our approach, we can forge a more resilient future, where health, sustainability, and progress converge.

This shift demands collective action – from individuals, communities, businesses, and policymakers. The stakes are high, but the potential rewards are transformative: healthier populations, restored ecosystems, and a food culture that nourishes both people and planet.

Homemade goodness! I made this pounded yam with goat meat and okra, inspired by my Ghanaian friend's recipe—reminds me of our own sadza! In Zimbabwe. It got me thinking, after such a meal, do we really know how much of what we have eaten nourishes our bodies and how much just turns to fat? The answer is definitely NO.

This is why we need the Mathematics of food.

References

1. https://www.wapt.com/article/deadly-chocolate-factory-explosion-highlights-combustion-risks-hazards-of-food-manufacturing/43446602#
2. https://www.fsis.usda.gov/recalls-alerts/elkhorn-valley-packing-recalls-boneless-beef-chuck-product-due-possible-e–coli-o103?permalink=4503898680325BB12B960C10BFCD86182C01002C205373614E0DC60D96DCD81C
3. https://www.cdc.gov/foodborneburden/index.html
4. https://revolutionized.com/how-can-we-address-food-safety-concerns-in-america/
5. https://www.canr.msu.edu/resources/keeping_foods_out_of_the_temperature_danger_zone_e3254
6. https://injuryfacts.nsc.org/work/work-overview/top-work-related-injury-causes/
7. https://www.osha.gov/ammonia-refrigeration
8. https://www.cdc.gov/nora/councils/hcsa/stopsticks/bloodborne.html

9. https://www.quincycompressor.com/streamline-food-beverage-manufacturing-process/
10. https://iotbusinessnews.com/2020/10/26/91544-how-predictive-maintenance-is-helping-businesses-save-millions-every-year/
11. https://pubmed.ncbi.nlm.nih.gov/31795665/
12. https://www.lung.org/clean-air/at-home/indoor-air-pollutants/cleaning-supplies-household-chem
13. https://www.bls.gov/opub/ted/2021/32-percent-of-nonfatal-injuries-resulting-in-days-away-from-work-treated-in-emergency-room-in-2019.htm

www.ingramcontent.com/pod-product-compliance
Lightning Source LLC
Chambersburg PA
CBHW041259210926
51847CB00009B/72